电池科普与环保③
Battery Science Popularization and Environmental Protection ③

电池大环保

Battery and Environmental Protection

（中英对照版）
(Chinese-English Version)

马建民 / 主编   *Edited by: Ma Jianmin*

咪柯文化 / 绘   *Illustrated by: Micco Culture*

廖　敏 / 译   *Translated by: Liao Min*

电子科技大学出版社
University of Electronic Science and Technology of China Press

·成都·

前言 Preface

人类对能源的探索永不停止

人类对能源的探索，从来就未曾停止。由于地球上可供开采的煤炭、石油、天然气等非再生能源十分有限，因此，现在全世界都将目光聚焦于太阳能、风能、核能、潮汐能等再生能源的开发与利用。

能源问题是关系国家安全、社会稳定和经济发展的重大战略问题。优化资源配置，提高能源的有效利用率，对于人类的生存和国家的发展都具有十分重要的意义。

如何积极发展新能源是人类必须共同面对的一项重大技术课题。新能源技术的不断进步，特别是动力系统的不断改进，为能源结构的转型提供了可能。然而，虽然新能源的类型很多，但世界上至今还没有实用的、经济有效的、大规模的直接储能方式。因此，人类还不得不借助间接的储能方式。

电能，作为支撑人类现代文明的二次能源，它既能满足大量生产、集中管理、自动化控制和远距离输送的需求，又具有使用方便、洁净环保、经济高效的特点。因此，电能可以替代其他能源，提高能源的利用效率。

我们今天所有的可移动电子设备，其运行都离不开电池。电池的出现使人类的生活更加便捷，特别是在信息时代来临之后，电池的重要性更为突出。我国不仅是世界排名第一的电池生产大国，也是世界排名第一的电池消费大国。

人类虽然在电池的研究方面已经取得了丰硕的成果，但还一直在寻找更好的电能储存介质。随着科学的发展、新能源技术的成熟，在未来，哪一种类型的电池能够脱颖而出还未可知。希望此书能激发孩子们对电池的兴趣，让他们在未来为我们揭晓谜底。

马建民

2024 年 3 月

Endless Exploration of Energy

Since ancient times, humans' quest for energy has never ceased. Given the limited reserves of non-renewable energy like coal, oil, and natural gas on Earth, the use of renewable energy like solar, wind, nuclear, and tidal power, has become the new global focus.

Energy is a major strategic issue that bears on national security, social stability and economic development. How to allocate and use energy in a better way means a lot for both individuals and countries.

How to develop new energy is a major technological topic facing humanity. With the development of new energy technology, especially the power system, energy structure transformation is made possible. However, despite various kinds of new energy, there is yet to be a practical, cost-effective, large-scale way of direct energy storage. Therefore, we have to resort to indirect methods to store energy.

Electricity, as a secondary energy propelling modern civilization, can support mass production, centralized management, automated control, and long-distance transmission. At the same time, electricity is clean, economical, efficient, environmentally friendly and easy to use. We can replace other energy with electricity to use energy better.

All the electronic mobile devices today can't operate without batteries. Batteries make our life more convenient. Its importance grows even more prominent with the advent of the Information Age. China is now the world's biggest producer and consumer of batteries.

Although we have gained so much in battery research, researchers are still looking for a better medium for power storage. With progress in science and new energy technologies, which type of battery will stand out still awaits our exploration. Hopefully, this book will spur children's interest in batteries and one day make them tell us the answer in the future.

Ma Jianmin

March 2024

废旧电池的危害及防治
Dangers of Used Batteries and Related Prevention and Treatment

有害元素 Harmful Elements	代表电池 Example	对人体的毒害作用 Toxic Effects on Human
镉 cadmium	镍镉电池 nickel-cadmium battery	镉进入人体后会使肝和肾受损，也会引起骨质松软，重者造成骨骼变形 Cadmium hurts our liver and kidneys, weakens and even deforms bones
铅 lead	铅酸电池 lead-acid battery	铅主要作用于人体的神经系统、循环系统、消化系统，能抑制血红蛋白的合成和代谢过程，还能直接作用于成熟的红细胞，对婴幼儿影响很大，能导致儿童体格发育迟缓，慢性铅中毒可导致儿童智力低下 Lead affects our nervous, circulatory and digestive systems. It can inhibit hemoglobin synthesis and metabolism, and harm mature red blood cells. The impacts are more serious for infants and toddlers. Lead may stunt the growth of children and long-term lead poisoning may cause intellectual disability

对环境的影响
Environmental Impact

当自然环境受到镉污染后，镉可通过生物体的富集作用，蓄积在生物体内，并通过食物链进入人体

Once the cadmium pollutes our nature, it can accumulate in organisms through biomagnification. When we eat, cadmium enters our bodies through the food chain

空气中的铅烟尘达到一定浓度后对人体是有害的。并且铅无法降解，排入环境很长时间之后仍然含有毒性

Airborne lead dust with certain concentrations is harmful to humans; lead does not degrade and remains toxic long after being released into the environment

防治措施
Prevention Measures

目前，人类对于镉、铅、钴这三种元素造成的环境污染的治理方法主要是利用植物修复法，即用绿色植物来转移、容纳或转化重金属，使其对环境无害。这种技术因其成本低，利于土壤生态系统的保持，对污染地景观有美学价值，且对环境基本没有破坏的，从而被大规模应用于净化土壤或水体中的污染物

Currently our way to prevent cadmium, lead, and cobalt pollution is phytoremediation, using plants to transfer, contain, or transform heavy metals to make them pollution-free. This cost-effective and environmentally-friendly technique can balance the soil ecosystems and make polluted landscapes beautiful, and thus it is widely applied to pollutants in soil and water

有害元素 Harmful Elements	代表电池 Example	对人体的毒害作用 Toxic Effects on Human
钴 cobalt	钴酸锂电池 lithium cobalt oxide batteries	当人体摄入钴超过 500 毫克时，就会发生钴中毒。钴中毒的临床表现为食欲不振、呕吐、腹泻等 When a person consumes more than 500 milligrams of cobalt, he/she will be poisoned. The clinical symptoms of cobalt poisoning include loss of appetite, vomiting, and diarrhea
氟 fluorine	锂离子电池 lithium-ion batteries	低浓度氟污染对人畜的危害主要为使牙齿和骨骼氟中毒。牙齿氟中毒表现为牙齿着色、发黄、牙质松脆、缺损或脱落。骨骼氟中毒表现为腰腿疼，骨关节固定、畸形，骨质密度增加，关节、韧带钙化等 Low levels of fluoride pollution can affect people and animals and cause fluorosis, which mainly harms their teeth and bones. In teeth, symptoms include discoloration, yellowing, brittleness, and even tooth loss. For bones, fluorosis leads to pain in the back and legs, stiff or deformed joints, increased bone density, and joints or ligaments calcification

对环境的影响 Environmental Impact	防治措施 Prevention Measures
钴在天然水体中常以水合氧化钴、碳酸钴的形式存在，或者沉淀在水底，或者被底质吸附，很少溶解于水中。土壤中的钴会严重影响周围动植物的生长和发育，土壤溶液中钴的浓度为 10 毫克/升时，可使农作物死亡 Cobalt exists in natural water as cobalt oxide or cobalt carbonate. It either settles at the bottom or gets absorbed by sediments and rarely dissolves in the water. Cobalt in soil seriously corrodes the nearby plants and animals. When its concentration reaches 10 mg/L, crops can die	植物修复法 Phytoremediation
氟污染可能使动、植物中毒，影响农业和畜牧业生产 Fluorine pollution can poison flora and fauna, disrupting agricultural and livestock production	为了防止过量氟对人体健康的危害，我国在《工业企业设计卫生标准》中规定了来自工厂的氟的排放量，并加以严格管控 To prevent excessive fluorine emission, China sets limits on fluoride emissions from factories in *Hygienic Standards for the Design of Industrial Enterprises* and enforces strict control measures

有害元素 Harmful Elements	代表电池 Example	对人体的毒害作用 Toxic Effects on Human
磷 phosphorus	磷酸铁锂电池 lithium iron phosphate battery	摄入过量磷，会对人体造成如骨质疏松、牙齿蛀蚀等损害。表现为影响钙的吸收、利用，对骨骼产生不良作用，但如果体内有足量的钙，可有效抵御高磷的不利影响 Too much phosphorus intake can lead to brittle bones and tooth decay. It can also stop your body from using calcium properly, which is bad for your bones. Try to take in enough calcium, it can protect your body from the harmful effects of too much phosphorus

对环境的影响
Environmental Impact

磷化工行业的主要污染物是废气、废水、固体废物。这些污染物中含有许多含磷化合物，它们会进入大气、江河湖海和陆地，是我国环境污染最主要的来源之一

The main pollutants from the phosphorus chemical industry are waste gases, wastewater, and solid waste. These pollutants contain many phosphorus compounds. When released into the atmosphere, rivers, lakes, seas, and land, these compounds become one of the major sources of environmental pollution in China

防治措施
Prevention Measures

1. 控制河流、湖泊水体磷污染，制定环境标准；
2. 加强立法管理，设立管理机构和加强水质监督管理；
3. 深层曝气、疏浚底泥，控制泥磷的释放；
4. 控制非点源磷污染负荷；
5. 种植水生植物，如水葫芦，可以吸收水中的含磷污染物质，从而降低水体的污染负荷；
6. 养殖草食性鱼类，以除掉水中大量的磷；
7. 对点源磷污染实施工程治理，限制含磷洗涤剂的使用，减少点源磷排放

1. Set rules to control phosphorus pollution in rivers and lakes, to keep water clean;
2. Make new laws and enhance water quality monitoring and management;
3. Perform deep layer aeration and dredge up silt at the bottom, so phosphorus sludge doesn't leak out;
4. Control phosphorus pollution from non-point sources;
5. Plant water-loving plants like water hyacinths and eichhornia crassipes. They can "eat" the phosphorus and clean the water;
6. Raise plant-eating fish to help remove excessive phosphorus from the water;
7. Start control projects to treat point-source phosphorus pollution, reduce the use of phosphorus-containing detergents, and limit phosphorus emissions

故事导读
Introduction

  电池王国是一个庞大的国度，其中生活着许许多多的电池家族，每个家族的电池人都有着特殊的本领。他们勤劳能干，驱动各种设备运转，促进人类科技不断发展。

  在电池王国，每天都有故事发生。自从锂锂继任国王以后，为电池王国的更新换代作出了卓越的贡献。偷偷潜入"历史回廊"的不速之客——镍霸，竟利用心灵宝石的力量，蛊惑了一大批废旧电池，组建起了一个名叫"废旧电池联盟"的组织，企图污染人类世界。

  面对这一场危机，锂锂将如何应对呢？一起来看看吧！

  The Battery Kingdom is a huge country, with many battery families living there. Each family's "battery men" have special skills. They are hard-working and capable, driving all kinds of equipment and promoting the continuous development of the human world.

  In the Battery Kingdom, there are stories every day. Since Lithium Li succeeded to the throne, he has made outstanding contributions to the renewal of the Battery Kingdom. The uninvited guest who sneaked into the "Historical Corridor" —Tyrant Nickel, used the power of the Mind Gem to seduce a large number of waste batteries and formed an organization called "Used Battery Alliance" in an attempt to pollute the human world.

  Faced with this crisis, how will LiLi respond? Let's take a look!

角色介绍
Characters

锂锂　Lithium Li

家族：锂离子电池

Family: Lithium-ion Battery

大铅　Big Lead

家族：铅酸蓄电池

Family: Lead-acid Battery

机器人 X　Robot X

闪闪 Shiny

身份：电池王国的守护精灵

Identity: Guardian Spirit of the Battery Kingdom

镍霸 Tyrant Nickel

家族：镍镉电池

Family: Nickel-cadmium Battery

目 录
Table of Contents

1 大战在即
The Great Battle Looms Ahead ... /001

2 废旧电池联盟
The Used Battery Alliance ... /007

3 解铃还须系铃人
The Cause, the Fix ... /015

电池大揭秘
Secrets behind Batteries /025

废旧电池的危害

The Hazards of Used Batteries……………………………………026

废旧电池资源化

Recycling of Used Batteries into Useful Resources……………035

不同类型电池的环保问题

Environmental Challenges of Different Batteries………………046

电池环保，个人应该怎么处理废旧电池？

How Should Ordinary People Handle Used Batteries?…………073

1

大战在即
The Great Battle Looms Ahead

收到机器人X的紧急通知，在电池王国与人类世界的交界处，有一个电池人村庄，被一支名为"废旧电池联盟"的军队侵占了。

于是，锂锂决定，马上前去平息战乱、调查原委。

Lithium Li received an urgent message from Robot X: at the border between the Battery Kingdom and the human world, a village of battery people had been taken over by a group calling themselves the "Used Battery Alliance".

Lithium Li decided to set out in no time to stamp out the chaos and find out what was going on.

收到最新消息，在本次侵略中，废旧电池反叛军无差别地攻击了当地电池居民，并掠夺土地资源与水资源。

Latest news: the Used Battery Rebels attacked every battery villager and raided their land and water resources.

看！这些电池人的状态，跟历史回廊里被镍霸用心灵宝石控制的电池人好像！

Look! These rebels seem just like the ones in the Historical Corridor—controlled by Tyrant Nickel with the Mind Stone!

恐怕这事又与镍霸有所关联……我们需要马上出发去制止他们！
It may be Tyrant Nickel again… We need to stop them now!

闪闪，你在吗？
Shiny?

有什么需要帮助的吗？
Yes, how can I help you?

我希望你能用空间宝石送我们到叛军所在的地方，我需要尽快调查清楚！
I want you to use the Space Stone and teleport us to the rebels; I need to find out the cause!

好的，陛下！
Yes, my lord!

不过，由于空间宝石被损坏，我无法将整个军团一起传送，只能分批次传送了……
But the Space Stone was damaged. I can't transport the legion altogether, but only in batches…

机器人X，马上召集王国护卫军！随时待命！
Robot X, gather the Royal Guards immediately! Let them wait for my order!

在闪闪的帮助下，锂锂与大铅带领第一批王国护卫军火速抵达了现场，将叛军团团包围起来。

With Shiny's help, Lithium Li and Big Lead led the first batch of Royal Guards to the scene. The rebels were soon hemmed in.

我向你们郑重承诺，放下武器，你们将被送往最好的地方疗养！

You have my word: lay down your weapons, and you'll be taken to the best place for care!

撒谎！我们一贯是被遗弃的，谁会在意我们的死活……

Liar! We are abandoned! No one cares about us…

正当王国护卫军与叛军僵持不下的时候，镍霸手持着心灵宝石，出现在众人面前！

Just as the two sides were at a deadlock, Tyrant Nickel appeared before everyone, holding the Mind Stone.

不要相信他！

Don't trust him!

我们镍镉电池家族曾为这个世界作出很多贡献，但一旦被人类利用完，还不是被扔到荒郊野外，更何况是你们！

We Nickel-Cadmium Batteries contributed so much to this world. But once humans used us, they just cast us away like rubbish. Not to mention you!

这……是真的吗？不要骗我们。

Is... is that true? Don't lie to us.

果然是你！镍霸！

It's you! Tyrant Nickel!

他手上有心灵宝石！大家后退！

He has the Mind Stone! Everyone, step back!

闪闪，请为我们打开空间保护罩！

Shiny, the space shield, now!

传送护卫军消耗了我大量的电力，现在，我的电量已经不足以使用残缺的空间宝石了！

Teleporting the army has drained too much of my power. I am not powerful enough to use the damaged Space Stone now!

005

锂锂担心王国护卫军由于无法得到空间保护罩的保护而被镍霸所控制，只能下令紧急撤离。

Lithium Li worried that without the space shield, the Royal Guards might fall under Tyrant Nickel's control. He had to order an emergency retreat.

糟了！王国护卫军，马上随我速速撤离！

Oh, no! Everyone, leave, Now!

嘿嘿，再过几天……这个世界……即将属于我了！

Heh, heh… In just a few more days… this world… will be mine!

2

废旧电池联盟
The Used Battery Alliance

赶走了前来平息叛乱的护卫军之后，镍霸加快了招兵买马的步伐。他利用心灵宝石的力量，快速在王国边境召集起一大帮被人类界定为有害的废旧电池。

就这样，"废旧电池联盟"迅速扩张，一个无比邪恶的计划正式开始……

After driving off the Royal Guards Lithium Li sent to crush them, Tyrant Nickel hurried to grow his army. With the power of the Mind Stone, he quickly gathered an army of used batteries, labeled as harmful by humans, at the kingdom's borders.

Thus, the "Used Battery Alliance" expanded quickly. A wicked plan was brewing…

同胞们！人类都称废旧电池污染环境，却不承认这都是他们自己造成的！

Fellow batteries! Humans blame us for polluting the environment, but they refuse to admit it is their own fault!

我们一旦没用了，就会被人类随意丢弃，他们不对我们进行绿色回收，最后还要给我们安上个"有害电池"的污名！

The moment we're no longer useful, they throw us away like trash! No recycling, just the smear of "hazardous batteries"!

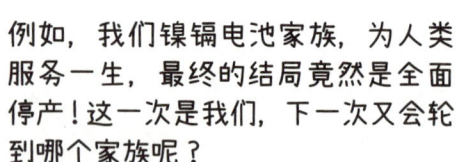

例如，我们镍镉电池家族，为人类服务一生，最终的结局竟然是全面停产！这一次是我们，下一次又会轮到哪个家族呢？

We Nickel-Cadmium Batteries dedicated all to the humans, and how did they thank us? By banning our production entirely! It was us this time—but who's next?

既然人类认为我们是威胁，惧怕我们，那我们何不将这份畏惧好好利用起来？

Human sees us as a threat: they fear us, so why not use it?

让我们联合起来，改变世界！我们要向遗弃我们的人类宣战，向电池王国争取我们应有的权利！

Let's unite and change the world! We will declare war on the humans for abandoning us and fight for the rights we deserve in the Battery Kingdom!

所有支持的人都举起手！

All those in favor, raise your hands!

支持！

I'm in!

支持！

Aye!

支持！

Aye!

我支持！

Aye!

支持！

Aye!

小贴士 Tips

对人体健康和生态环境危害较大、被列入《国家危险废物名录》的废旧电池主要有含汞电池、含镉电池以及铅酸蓄电池。

Batteries that pose serious risks to human health and the environment—such as mercury-containing batteries, cadmium-containing batteries, and lead-acid batteries—are listed in *National Catalogue of Hazardous Wastes*.

另一头，电池王国内，一座座名为"退役电池家园"的疗养院已经建设完工。这样一来，电池人即使在退役之后，也能有个安身之所，并继续为电池王国发光发热。

Meanwhile, in the Battery Kingdom, more and more "Home of Retired Battery" had been built. These places would assure retired batteries a safe life after retirement. They could rest while still doing their share for the kingdom.

请各位电池人——**排队报名**
Attention, all batteries!
Line up to register!

太好了，我报名！
Awesome! I'm signing up!

退役生活有了保障，真是一个好消息啊！
No worry after retirement—what great news!

我也报名！
Count me in, too!

安置废旧电池的地方是建设好了，但后续需要解决的问题还有许多……

正在锂锂进一步思考解决方案时，他又收到了来自机器人 X 的一则紧急通知……

The homes for retired batteries were built, but many problems were still left to be solved…

Just as Lithium Li was looking for a solution, an urgent message arrived from Robot X…

退役电池家园
Home of Retired Battery

我好像知道……前任国王所说的"危机"是指什么了……

I think I finally understand… what the former king meant by the crisis…

目前，我能为他们做的，只有这些了……

For now, this is all I can do for them…

你当上国王以后，能力真是一天比一天强了！把电池王国建设得一天比一天好！

Ever since you became the king, you've become more mature. Our kingdom is so much better now!

紧急通知！紧急通知！
Urgent Alert! Urgent Alert!

> 头好疼……
> My head hurts…

> 哇，恶心……哕！
> Ugh, I feel sick… gag!

> 好热！
> So hot!

根据全球医疗系统统计，最近一年来，人们患上癌症的概率大幅度上升，专家正在就未来的患癌概率进行研判。

According to the statistics by global healthcare system, the probability of developing cancer for people has greatly increased over the past year. Experts are currently analyzing the trends of future cancer risk.

小贴士 Tips

废旧电池被腐蚀后的颗粒不仅能渗透土壤，还能随风飘散在空气中。

The particles from corroded used batteries can seep into the soil and spread through the air with the wind.

原来，仅仅一年时间，人类世界就爆发了大规模电池污染，医院已经人满为患。

锂锂想，人类世界才是废旧电池的源头，只有让人类都树立起环保意识，才能应对这场危机。

In just one year, the human world was hit by a massive battery pollution, and now the hospitals were packed full!

Lithium Li thought to himself, "Humans are the real source of all these discarded batteries! If we can help them understand the importance of protecting the environment, maybe we can solve this problem!"

可恶！我们安置了大部分的废旧电池，没想到他们竟然选择攻击人类！

Awful! We gave homes to so many used batteries, but they chose to attack humans instead!

要想彻底解决废旧电池的问题，必须要与人类共同努力！

To fix this problem once and for all, we need to work together with humans!

闪闪，你能帮助我与人类世界沟通吗？

Shiny, can you help me talk to the human world?

此时，镍霸正与"废旧电池联盟"的成员们在一起开会，成员们纷纷向他汇报着自己的"复仇成果"。

Meanwhile, Tyrant Nickel was in a meeting with the alliance members who were bragging about their "revenge fruits".

嘿嘿！我们最近将已经牺牲的联盟成员投放到了人类的农田里。

Hehe! We've recently put the sacrificed members into human farmlands.

毒素将会慢慢渗透到土壤里，他们种的庄稼将会慢慢受到影响，最终颗粒无收！

The toxins will slowly sneak into the soil, and ruin the crops!

我把自己体内的有毒物质投到了水里。喝了有毒的水，他们将会难以生存！

I dumped the toxic substances from my body into the water. Those who drink it won't survive!

很好！继续努力！新世界的规则，将由我们来制定！

Great! Keep it up! We'll set the rules for the new world!

014

2

解铃还须系铃人
The Cause, the Fix

在人类世界，不少人都见识到了现代工业污染对大自然的影响，并开始对废旧电池造成的污染有所重视。

In the human world, many people see what modern industrial pollution has done to nature. They started to pay attention to the pollution caused by used batteries.

我好怀念小时候，这里还是绿水青山！

I miss the old days! The hills were green and the water was clear!

好多化工原料和废旧电池不经处理，就被随意倾倒和填埋，大自然无法"消化"，自然会影响到人类的生存！

Well, too many chemicals and used batteries are dumped and buried without proper treatment. Our mother nature can't "digest" them, and surely it will affect our lives!

是呀，难怪现代人总是生一些奇奇怪怪的病！
Yeah, no wonder people these days get all those strange diseases!
听说，乱扔电池会被罚款！
Oh, right, and people will be fined for littering batteries!

正当人类因为环境污染带来的健康危机而感到焦头烂额时，人类世界所有的电子设备都同时收到了来自电池王国的讯息——"请与电池王国一起努力！改善废旧电池的生存环境！"

Just as humans were overwhelmed by the health crises caused by environmental pollution, every electronic device in the human world received a message from the Battery Kingdom—"Work with the Battery Kingdom to improve the living environment of used batteries!"

Beep— 滴— Beep— 滴— Beep—

请与电池王国一起努力！
改善废旧电池的生存环境！
——电池王国

Work with the Battery Kingdom to improve the living environment of used batteries! —The Battery Kingdom

请与电池王国一起努力！
改善废旧电池生存环境！
——电池王国

都——

耶！成功了！人类收到讯息了！
We made it! Humans got our messages!
万众一心！众志成城！
Together we can do anything !

021

人类收到信息后，决定加大对废旧电池集中回收的力度。专业的回收公司开始集中出现，这一切都昭示着废旧电池的回收处理问题正在受到重视。

After getting the messages, humans decided to step up their efforts to collect used batteries. Recycling companies popped up everywhere. Good! People started to care about recycling and disposing the used batteries!

电池需要专门回收，放这里！
Batteries need to be recycled—put them here!

废旧电池专门回收处
Used Battery Collection Spot

废旧电池回收加工厂
Used Battery Recycling Plant

含重金属元素的废旧电池
Used Batteries with Heavy Metals

无害废旧电池
Harmless Used Batteries

含重金属元素的废旧电池处理设备
Disposal Equipment for Used Batteries with Heavy Metals

已拆解的含重金属元素废旧电池
Disassembled Used Batteries with Heavy Metals

回收废旧电池有利于环境保护,尤其是含重金属元素的电池。一定要仔细分拣,进行专业化处理!

Recycling used batteries helps protect the environment, especially those with heavy metals. We should sort them out and handle them properly!

随着人类对废旧电池回收力度的增强，"废旧电池联盟"被电池王国与人类联手瓦解了，他们原先安扎的营寨被人类改造成为了电池回收工厂。

就这样，镍霸的邪恶计划彻底落空。不过，为了达到自己的目的，他可不会轻易放弃……

As humans worked harder to recycle used batteries, the "Used Battery Alliance" was dismantled through a collaboration between the Battery Kingdom and humanity. Their former camps were transformed into battery recycling factories.

Tyrant Nickel's wicked plan completely fell apart. But, was he to give up?

废旧电池回收工厂
Used Battery Recycling Plant

这里的电池人都被人类抓走了！
Our army was captured by humans!

可恶！咱们走着瞧！我还会回来的……
No! We'll see! I'll be back...

电池大揭秘
Secrets behind Batteries

废旧电池的危害
The Hazards of Used Batteries

在我们的生活中，随着电子产品、家电种类的丰富，废旧电池的数量和种类也在不断增加，再加上新能源汽车的普及，汽车上搭载的动力电池数量也在增长。

据中国电池工业协会统计，我国每年消费的普通电池就多达 80 亿只。

As we use more and more electronic devices and appliances, the numbers and types of used batteries are also growing. Also, with the rise of electric cars, the number of batteries embedded in cars is increasing too.

According to the China Battery Industry Association, people in China consume 8 billion regular batteries every year.

> 退役之后，我们将去哪里呢？
> What will happen to us when we finish our work?

电池属于消耗品，废旧电池时刻都会产生。电池电量耗尽后，就成了废旧电池。

作为人类制造的工业产品，废旧电池能无害地回归大自然吗？

Batteries are consumables, so there will always be used batteries. Once the battery runs out of power, it becomes a used battery.

As an industrial product made by humans, can a used battery dispose on its own and merge into the environment harmlessly?

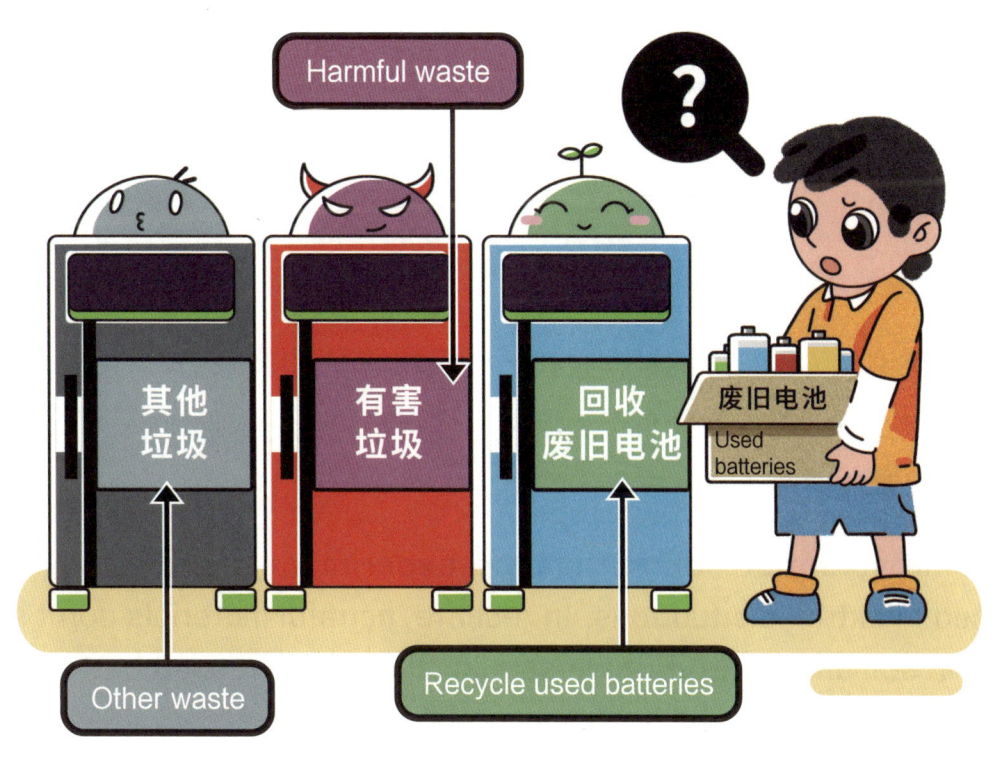

小测试 Quiz

废旧电池该放在哪里呢？
Where should we put used batteries?

- A. 标有"有害垃圾"的垃圾桶　　A. Trash bin marked "Harmful Waste"
- B. 标有"其他垃圾"的垃圾桶　　B. Trash bin marked "Other Waste"
- C. 社区街道　　　　　　　　　C. Community Street
- D. 标有"回收废旧电池"的垃圾桶　D. Trash bin marked "Recycle Used Batteries"

Harmful waste

其他垃圾 / 有害垃圾 / 回收废旧电池 / 废旧电池 Used batteries

Other waste　　Recycle used batteries

废旧电池对环境的危害实验
Experiment: The Hazards of Used Batteries to the Environment

将一粒纽扣电池扔到水里，可使 600 吨水受到污染。

If you throw a small button cell into the water, it can pollute 600 tons of water!

一节一号电池，可永久污染一平方米的土地。

A single D-size battery can permanently pollute one square meter of land.

一节含汞电池可使 60 立方米大气汞超标。

A mercury battery can make 60 cubic meters of air polluted with mercury.

有科学家认为，上面的数据是在一些极端情况下得出的实验结果，在现实中，有害物质不可能均匀分布于土壤和水体里。但是，人们应该承认，含有重金属的废旧电池会对环境造成严重污染。

Some scientists think that the experiment results above were obtained in extreme situations. In real life, harmful materials don't spread evenly in soil and water. But we should all agree that used batteries with heavy metals can hurt the environment badly.

电池在正常使用过程中，其组成物质被封存在电池外壳内部，并不会对环境造成影响。但当被填埋处理一个月左右，其金属外壳就会被腐蚀、穿孔，内部的重金属和酸、碱等随即泄漏出来，废旧电池中的有害物质就会进入土壤、水体，对环境造成污染。

When a battery is used the right way, the parts of the battery stay safely inside it and won't harm the environment. But if it's buried in the ground for about a month, the metal outside can get rusted and pierced. Then, the harmful materials inside, like some heavy metals, acids, and alkalines, will leak out and then pollute the soil and water.

废旧电池　used batteries

分解　break down

融入土壤　mix into the soil

人一旦食用受污染的由土地生产的农作物或是喝了受污染的水，这些有毒的重金属就会进入人的体内，慢慢地沉积下来，对人类健康造成极大的威胁！

When people eat crops grown on polluted land or drink polluted water, the harmful metals can get into their bodies. Over time, these harmful materials will build up and can be very dangerous to our health!

1939年11月9日，日本神奈川县某脑科医院接收了一名患病的男子。

该男子一开始的症状只是面部浮肿，仅仅3天时间，浮肿便从面部蔓延到了脚部。到了第8天，他的视力开始减退，并伴随自言自语、不断哭泣的表现。后来甚至发展为神志不清，人们都认为他"疯了"，于是将他送入了医院。最终，这名男子在痛苦中，因心力衰竭而死亡。

On November 9, 1939, a man was admitted to a brain hospital in Kanagawa Prefecture, Japan.

At first, the man's only symptom was face swelling. Within just three days, the swelling spread from his face to his feet. By the eighth day, he began to lose his vision, and he started talking to himself and couldn't stop crying. Eventually, his condition worsened, and he became delirious. People believe that he had gone "mad". He was then taken to the hospital. Finally, the man passed away in pain due to heart failure.

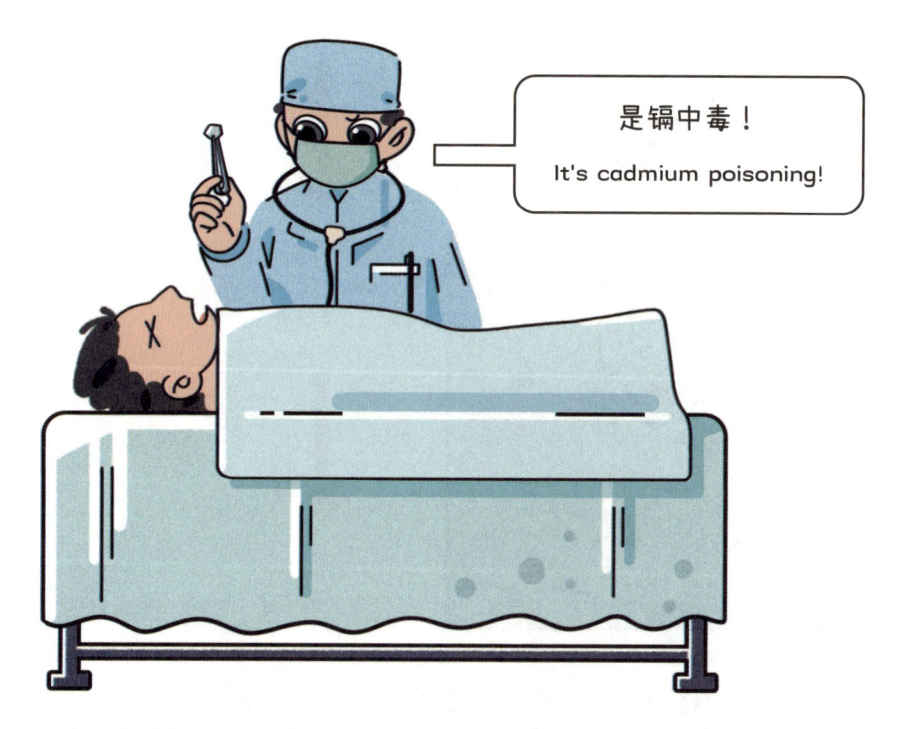

自此以后，与其同村居住的人中又陆续出现了15名同样症状的人。这引起了医学研究人员的注意。

Soon, 15 more people from his village started showing the same strange symptoms. This caught the attention of medical researchers.

医学研究人员经过调查后了解，原来他们都喝了一个商店旁边的一口井里的井水。这井水有什么特别之处呢？

原来，在距这口水井 5 米深的土里竟埋着几百节废旧电池！废旧电池在被腐蚀后，泄漏出的有害物质污染了井水，导致了这一场悲剧。

After investigating, they found out that all the people had been drinking water from a well next to a store. So, what was special about this well?

It turned out that just 5 meters below from the well, there were hundreds of used batteries buried in the ground! As the batteries were corroded, harmful materials leaked out and polluted the water, and caused this sad story.

井下 5 米
5 meters below the well...

啊，是废旧电池！
Ah, it's used batteries!

电池小知识
Battery tip

镉对人体的主要危害是肾毒性，还会续发"痛痛病"（引起骨质疏松、软骨病和骨折），同时还是致癌物质。

Cadmium is harmful to the human body, mainly damaging the kidneys. It can also cause "Itai-Itai Disease" — a series of bone diseases like osteoporosis, vitamin D deficiency, and fracture. Besides, cadmium can also lead to cancers.

废旧电池中的有害物质不仅仅是通过土地传播，在海水中时，它们的影响范围可以更广。

But did you know that harmful materials from used batteries don't just spread through the soil? When used batteries are thrown into the sea, the damage is even bigger.

最初，人们对废旧电池污染环境的认识并不深刻，在 2006 年之前，渔船上的人不时会将废旧干电池随手扔进海里。

A long time ago, people didn't know much about how harmful used batteries could be. Before 2006, fishermen sometimes threw used dry cells into the sea.

科学家发现，生活在泡过废旧电池的海水里的鱼和虾都会中毒，解剖后发现，它们的体内都含有重金属，如果被人吃了，后果不堪设想。

Scientists found that in the seawater polluted by used batteries, fish and shrimps became poisoned. When they looked inside these animals, they found dangerous metals. If humans eat these fish or shrimps, there will be serious consequences.

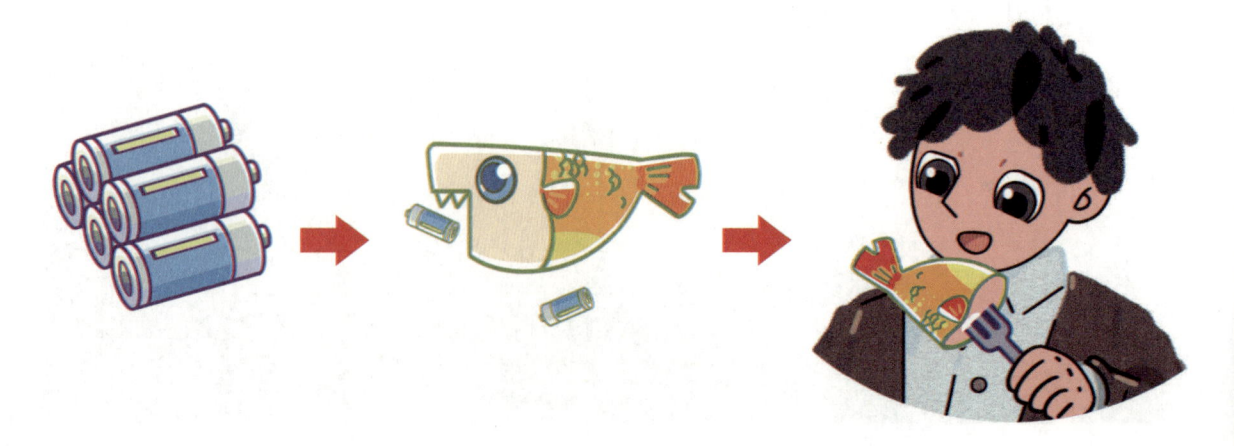

那么，不含重金属的电池是否对环境就完全友好呢？那可不一定，即使是号称"绿色电池"的锂离子电池和不含汞的干电池，它们的电解质溶液也会造成土壤或者水体污染。

Now, you might think that batteries without harmful metals are safe. But that's not always true. Even lithium-ion batteries, also called "green batteries", and dry cells without mercury can still pollute the soil and water because of the chemicals inside them.

干电池电解质粉末
powder in dry cells

一周后
one week later...

研究人员做了这样一个小实验，将干电池电解质粉末倒入盛开着鲜花的花盆内，并保持一个适宜其生长的环境，仅仅一个星期，原本生机勃勃的鲜花枯萎了。

Researchers did a little experiment. They poured the powder from dry cells into a flowerpot with a blooming flower and kept the environment just right for the flower to grow. But after just one week, the flower, once full of life before, had withered and died.

废旧电池里的汞、镉、铅都是重金属，重金属污染的最大特点是它们在自然界不能降解，只能迁移。也就是说，一旦水和土壤被污染，污染就无法消除。且这些重金属容易在生物体内蓄积，随着时间推移，会发生畸变，最终导致生物死亡。

The mercury, cadmium, and lead in used batteries are all heavy metals. The biggest problem with heavy metal pollution is that it can't break down in nature. Once water and soil are polluted, it can't be cleaned up. These heavy metals can build up in animals and plants. Over time, they can cause harm and even make living things sick or die.

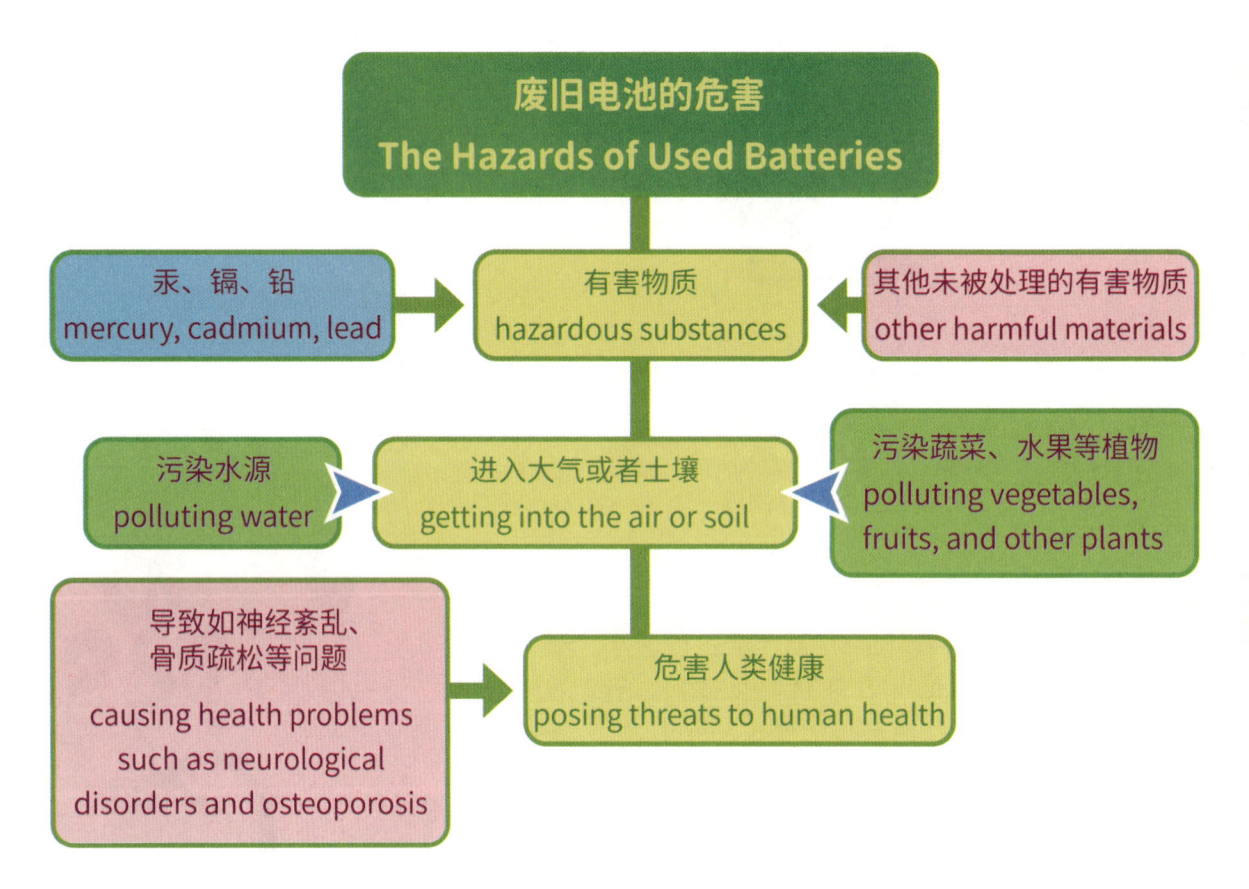

废旧电池的危害
The Hazards of Used Batteries

- 汞、镉、铅 mercury, cadmium, lead → 有害物质 hazardous substances ← 其他未被处理的有害物质 other harmful materials
- 污染水源 polluting water ← 进入大气或者土壤 getting into the air or soil → 污染蔬菜、水果等植物 polluting vegetables, fruits, and other plants
- 导致如神经紊乱、骨质疏松等问题 causing health problems such as neurological disorders and osteoporosis → 危害人类健康 posing threats to human health

电池大家族虽然对人类的生产和生活有卓越的贡献，但是其对生态环境造成的威胁，也渐渐地引起人们的忧虑。

Battery families have been very helpful to people in many ways, but they can also be bad for the environment. People are starting to worry about how they affect nature.

废旧电池资源化
Recycling Used Batteries into Useful Resources

有专家提出，我们不需要太过担心废旧电池，因为普通的电池外面有不锈钢或碳钢作为外包皮，里面的组成物质只要不泄漏出来，就不会造成环境污染。

当然，也有人持反对意见，认为废旧电池带来的污染触目惊心，后果非常严重。

Some experts say that we don't need to worry too much about used batteries, because they are wrapped in a metal casing made of stainless steel or carbon steel. As long as the materials inside don't leak out, these used batteries won't pollute the environment.

However, others disagree and believe that the pollution caused by used batteries is alarming, and the outcome can be very serious.

电池可以回收，不会污染环境。

Batteries can be recycled and won't pollute the environment.

如果随便乱丢电池，会污染土地和水。

But if people throw them away carelessly, they could pollute the land and water.

废旧电池回收处

Used Battery Recycling Spot

事实上，以上两种观点，都有一定道理。

部分电池中确实含有对人体有害的重金属，然而在如今的科学技术水平之下，废旧电池中95%的物质均可以回收再利用。尤其是作为不可再生资源的重金属，它们的回收价值很高。

但如果人们不提升环保意识，不重视生活垃圾的收集、分类及处理，将大量废旧电池与普通生活垃圾混放、填埋，仍然会导致重金属泄漏，进而污染土壤和地下水。这样一来，不仅会造成环境污染，对资源也是一种浪费。

In fact, both of the views above make sense.

Some batteries do contain harmful heavy metals that can be dangerous to people. But with today's science and technology, 95% of the materials in used batteries can be recycled. This is especially important for heavy metals, because they are non-renewable and have high recycling value.

However, if people do not raise their own environmental awareness and fail to properly collect, sort, and dispose of household refuse, and if they mix large amounts of used batteries with household refuse and bury them, heavy metals could leak out. This will pollute the soil and groundwater. In this case, the environment will be polluted and valuable resources will be wasted.

针对废电池的回收和处理，有人总结出了一些科学的程序及方法，一起来了解一下吧！

Experts have come up with smart ways to recycle and process used batteries. Let's learn about them together!

废旧电池的回收方式
Ways to Recycle Used Batteries

随着人们环保意识的提高、完善有效的回收网络和体系的建立，废旧电池造成的污染问题将会逐渐被解决。

目前，废旧电池的回收方式主要包括梯次利用和拆解再生。

As people care more about the environment and better recycling systems are built, the pollution from used batteries will slowly be fixed.

Currently, the main ways of recycling used batteries are Cascade Utilization, and Breakdown and Recycling.

梯次利用　Cascade Utilization

梯次利用，是指将已经使用过的、且已达到原设计寿命的电池，通过技术处理使其功能全部或部分恢复，再继续投入使用的过程，该过程属于基本同级或降级应用的方式。

例如，动力电池的实际容量降低到80%后，不再适用于电动汽车，但仍可以用于对功率和耐久性要求较低的其他领域，如低速电动汽车、通信基站和储能设备等。

Cascaded utilization refers to the process of reusing batteries that have reached the end of their designed life. Through technical treatment, their functions are either fully or partially restored, and they are then put into further use, typically for applications of equal or lower grade.

For example, when a power battery's actual capacity drops to 80% and is no longer suitable for use in eletric cars, it can still be used in other areas with lower power and durability requirements, such as low-speed electric cars, communication stations, and energy storage devices.

梯次利用流程图
Cascade Utilization Flowchart

电池耗尽
battery exhaustion

回收
collection

技术处理
technical treatment

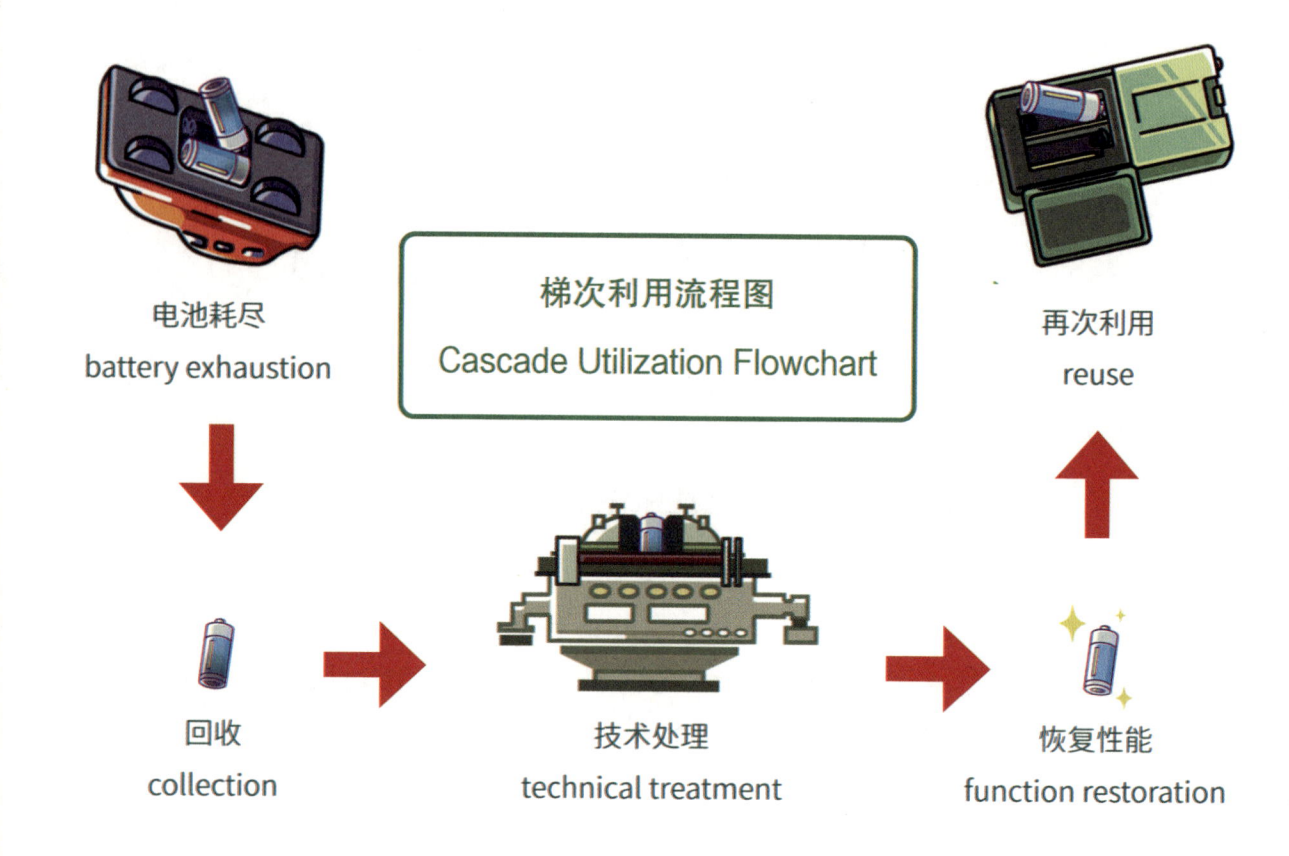

恢复性能
function restoration

再次利用
reuse

拆解再生　Breakdown and Recycling

拆解再生，是指电池彻底报废后进行拆卸、破碎、分选、材料修复或熔炼，在此过程当中，提取有价值的材料。

例如废旧锂离子电池的拆解回收，主要是将回收的废旧锂离子电池放电后，利用锂电池回收处理设备，使正、负极极板上的活性材料与铜铝箔得到有效分离，将电池处理为有价值的物料。

Breakdown and recycling refers to the process of breaking down, crushing, sorting, material restoration, or smelting the battery after it is completely used up. In this process, valuable materials are extracted.

Take the breakdown and recycling of used lithium-ion batteries as an example. After discharging the batteries, special recycling machines are used to separate the active materials on the positive and negative plates from the copper and aluminum foils, turning the battery into valuable materials.

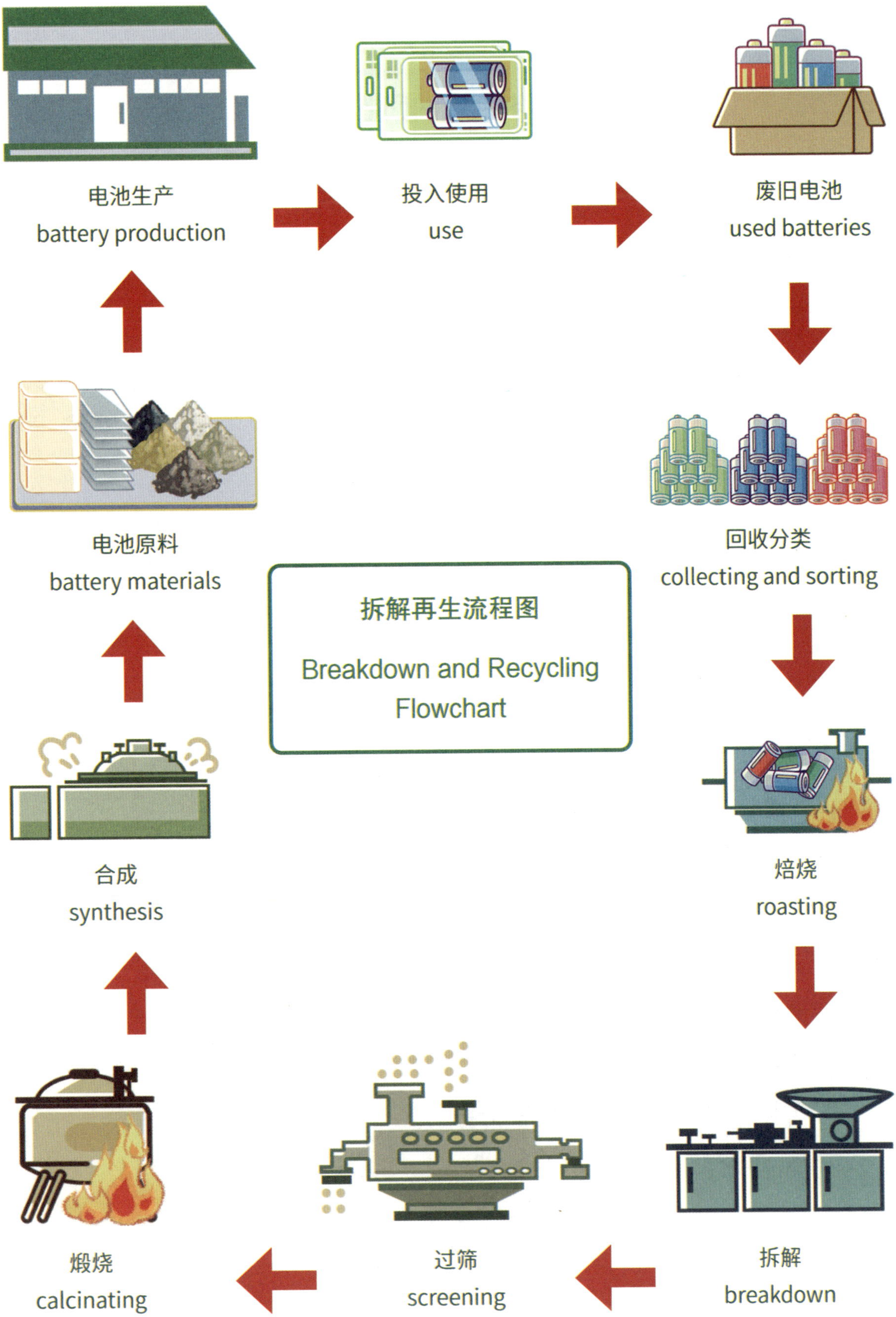

电池生产 battery production → 投入使用 use → 废旧电池 used batteries → 回收分类 collecting and sorting → 焙烧 roasting → 拆解 breakdown → 过筛 screening → 煅烧 calcinating → 合成 synthesis → 电池原料 battery materials → 电池生产 battery production

拆解再生流程图
Breakdown and Recycling Flowchart

039

科学合理的回收程序
Scientific Recycling Process

在废旧电池的资源再利用过程中，加工处理只是一方面，更重要的是从源头杜绝污染，做好回收工作。

During the recycling of used batteries, processing is just one aspect. More importantly, pollution should be prevented from the source, and proper recycling should be carried out.

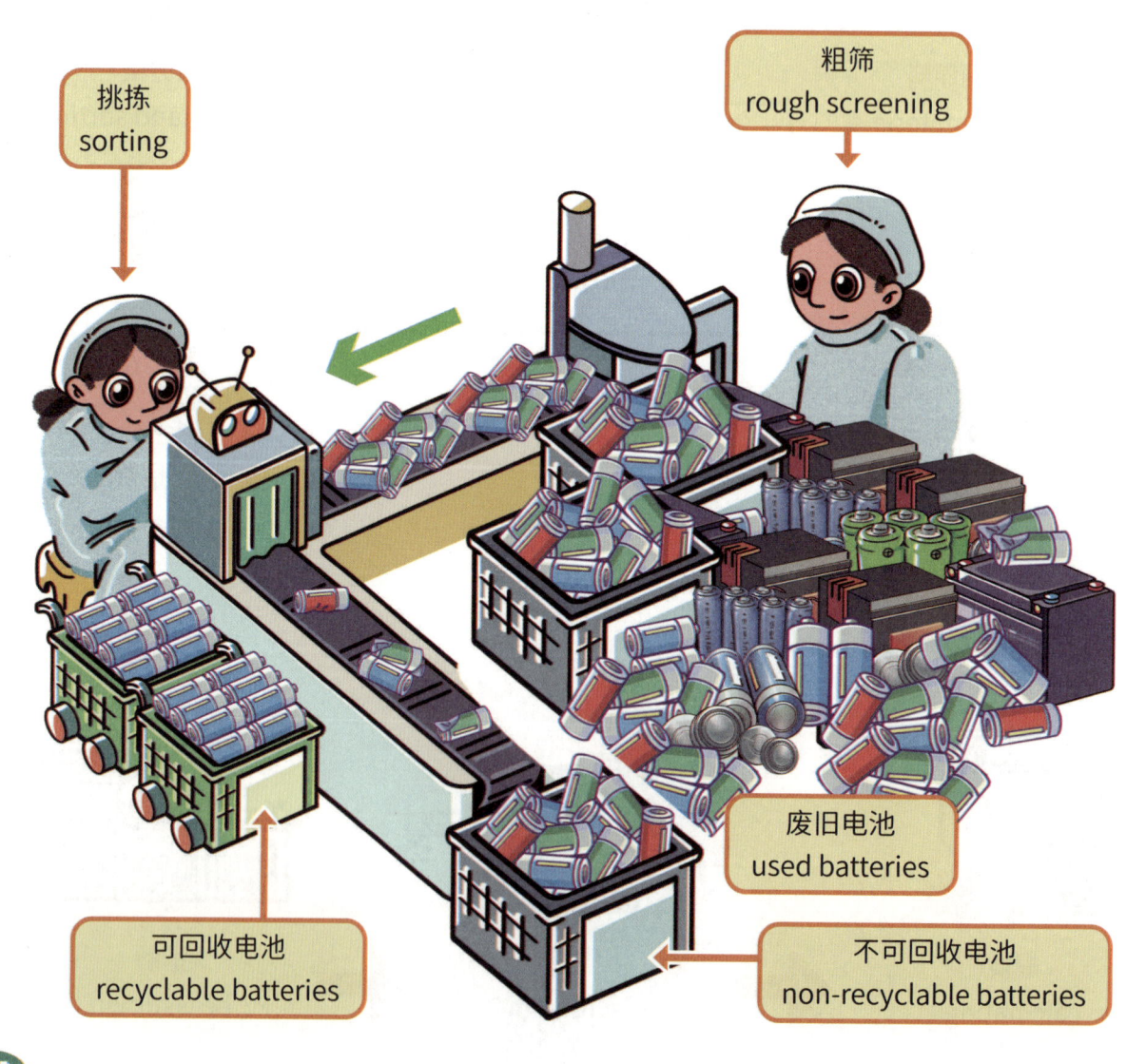

- 挑拣 sorting
- 粗筛 rough screening
- 废旧电池 used batteries
- 可回收电池 recyclable batteries
- 不可回收电池 non-recyclable batteries

如果不严格执行分类回收，必将会造成"集中生产，分散污染；短期使用，长期污染"的局面。

废旧电池大量回收的工作必须要完成，那么，专业的公司如何科学合理地完成回收？回收再利用的过程是否井然有序？下面这些操作步骤很关键，一起来看看吧！

If we don't strictly follow recycling and sorting rules, it will lead to a situatuin of "centralized production, scattered pollution; short-term use, long-term pollution".

Since the large-scale recycling of used batteries must be done, how can professional companies carry out recycling smartly? Is recycling organized? The following steps are super important. Let's check them out!

① 放置专用的废旧电池回收桶
place special recycling bins

② 专人定期上门收集
arrange regular home visits for collection by dedicated staff

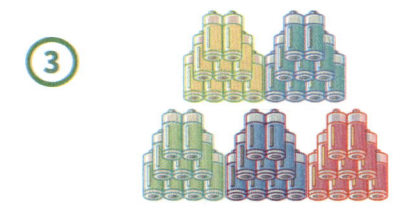

③ 电池分类
sort the batteries

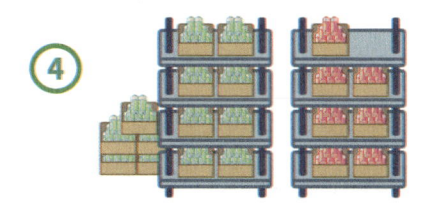

④ 在库房中分类并安全储存
sort and store used batteries safely in warehouses

⑤ 集中收集一定数量后，运至专门的处理工厂
send them to specialized treatment plants after gathering a certain amount

⑥ 处理回收稀有重金属
process and recycle rare heavy metals

废旧电池的处理方法
Ways to Process Used Batteries

在国际上，废旧电池的通用处理方式大致有三种，分别是固化深埋、存放于废矿井中和回收利用。

Internationally, there are three main ways to process used batteries: solidification and deep burial, storage in waste mines, and recycling.

固化深埋、存放于废矿井中
Solidification and Deep Burial, Storage in Waste Mines

在早期，废电池一般都被运往专门的有毒、有害垃圾填埋场。这种做法不仅要花费大量人力物力，而且还浪费了不少可作原料的有用物质，可以说是最不可取的处理方式。

In the past, used batteries were usually sent to landfills for toxic and harmful waste. This way wastes a lot of manpower and resources. Also, it wastes a lot of useful materials, because they can be used as raw materials. Thus, this way is considered the least desirable treatment.

回收利用　Recycling

经过工业处理之后的废旧电池，所含的原材料可以被有效地提取，经过技术加工后，最终可实现再次利用。

After industrial treatment, the raw materials in used batteries can be extracted and reused through further industrial processing.

废旧电池工业处理技术
Industrial Treatment Techniques for Used Batteries

常压热处理　Atmospheric Heat Treatment

用常压热处理在处理废旧电池时，通常有以下两种方法。

（1）在低温下加热旧电池，可提取挥发出的汞，在温度更高时可回收锌和其他重金属。

（2）在高温下焙烧废旧电池，使其中易挥发的金属及其氧化物挥发，残留物可作为冶金中间物产品或另行处理。

There are typically two methods used for the atmospheric heat treatment of used batteries.

(1) Heat the used batteries at a low temperature to extract mercury. At higher temperatures, zinc and other heavy metals can be recovered.

(2) Roast used batteries at high temperatures to vaporize volatile metals and oxides. The residue can be used as metallurgical intermediate products or be further processed.

真空热处理　Vacuum Heat Treatment

真空热处理的过程：首先，在废旧电池中分拣出镍镉电池，将其在真空中加热，其中的汞迅速蒸发，可将其回收，然后将剩余材料磨碎，用磁体提取金属铁，再从余下粉末中提取镍和锰。

The process of vacuum heat treatment: first, sort out nickel-cadmium batteries from the used batteries, then heat them in a vacuum, causing the mercury to evaporate rapidly for recovery. The remaining materials are then ground, and metallic iron is extracted using a magnet. Finally, extract nickel and manganese from the remaining powder.

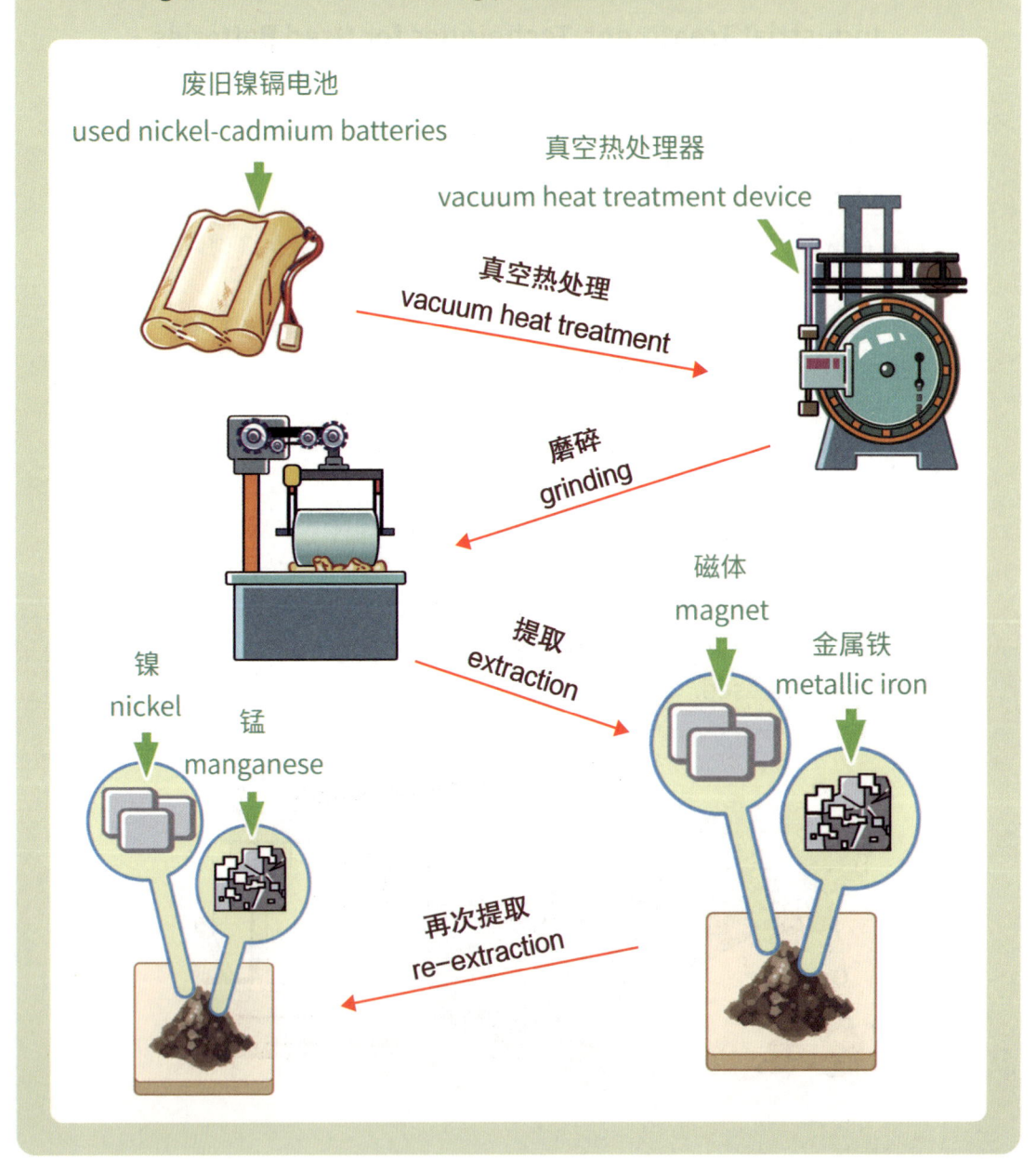

废旧镍镉电池
used nickel-cadmium batteries

真空热处理器
vacuum heat treatment device

真空热处理
vacuum heat treatment

磨碎
grinding

磁体
magnet

金属铁
metallic iron

镍
nickel

锰
manganese

提取
extraction

再次提取
re-extraction

湿处理　Wet Processing

将各类电池（除铅酸蓄电池外）均溶解于硫酸，然后借助离子树脂从溶液中提取各种金属。

All types of batteries (except lead-acid batteries) are dissolved in sulfuric acid, and then ion exchange resins are used to extract various metals from the solution.

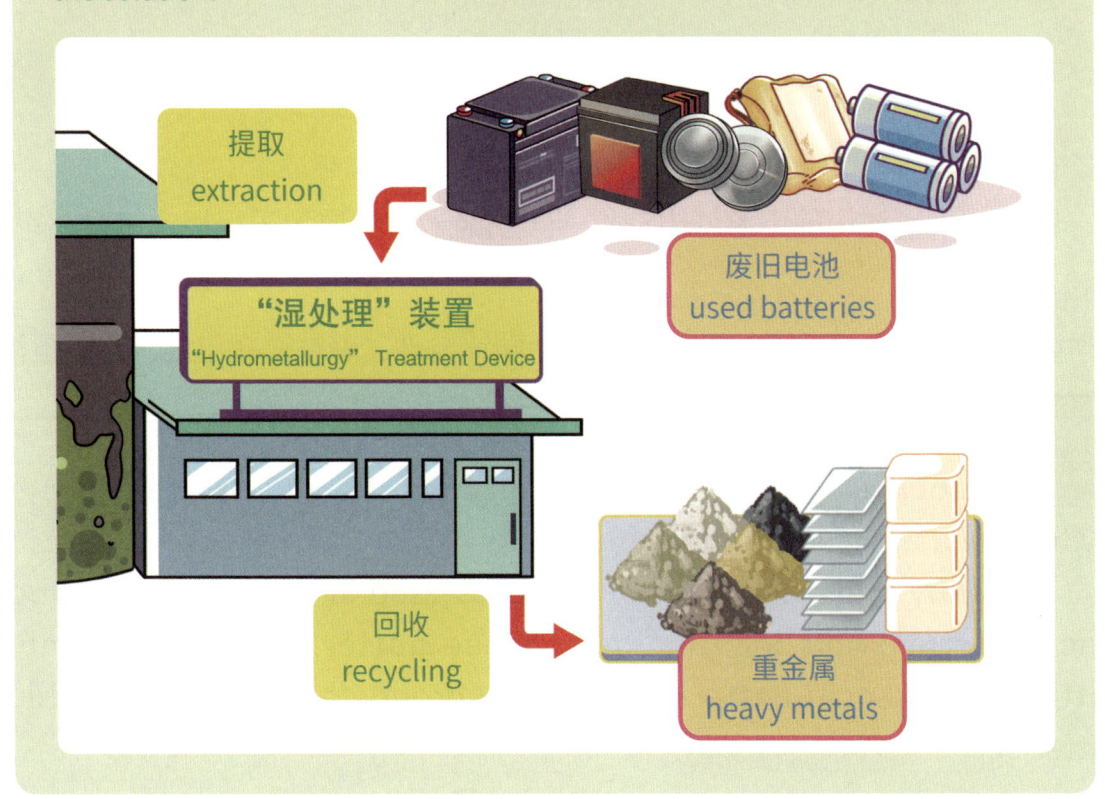

提取
extraction

废旧电池
used batteries

"湿处理"装置
"Hydrometallurgy" Treatment Device

回收
recycling

重金属
heavy metals

热处理法的成本较高，由于是高温提取，获取的贵重金属不够纯净，能源消耗也比较大。

相比于热处理法，湿处理法获得的贵重金属纯度很高，能提取出电池中所包含的 95% 的物质，而且可以省去分拣环节，处理成本也有所降低。

The cost of heat treatment is higher. And the precious metals obtained through high-temperature extraction are not pure enough. This way also consumes a lot of energy.

In comparison, hydrometallurgy treatment produces precious metals with higher purity. This method can extract 95% of the materials in the batteries. It can also skip the sorting and reduces the treatment cost.

不同类型电池的环保问题
Environmental Challenges of Different Batteries

作为提供直流电的电源，电池在今天已经被广泛应用于民用、工业和军工等诸多领域。

目前，我国主要应用的电池种类包括碳锌电池、碱性锌锰电池、镍镉电池、铅酸蓄电池、镍氢电池、锂电池等。

As a source of direct current, batteries are widely used today in civilian, industrial, and military fields.

Currently, the main types of batteries used in China include carbon-zinc batteries, alkaline zinc-manganese batteries, nickel-cadmium batteries, lead-acid batteries, nickel-hydrogen batteries, and lithium batteries.

常见干电池的环保问题
Environmental Issues with Common Dry Cells

过去，在日常生活中，人们最常接触到的干电池是碳锌干电池，它因正、负极的反应物质分别为二氧化锰和锌而得名。

但人们在使用过程中发现，碳锌干电池存在液体外溢的现象。

In the past, the most common dry cell people used was the carbon-zinc dry cell. It is named for its use of manganese dioxide and zinc as the positive and negative materials.

However, people discovered that carbon-zinc dry cells leaked "liquid" during use.

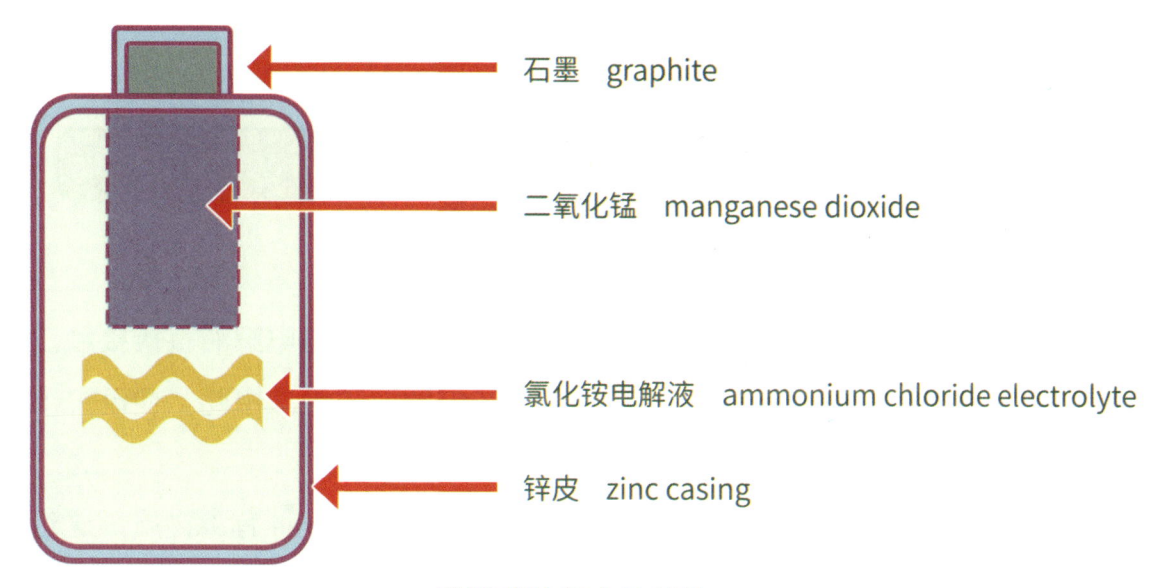

石墨　graphite

二氧化锰　manganese dioxide

氯化铵电解液　ammonium chloride electrolyte

锌皮　zinc casing

早期碳锌干电池结构
Structure of Early Carbon-Zinc Dry Cell

原来，这种漏液的情况是因为电池外壳——锌，和氯化铵电解液发生了反应，导致外壳破碎。

This leakage happens because the zinc casing reacts with the ammonium chloride electrolyte inside, causing holes in the casing.

所以在那个时候，当电池电量耗尽后，需要及时将其从电器内取出来，以免腐蚀电器。

后来，人们发现碳锌干电池不仅有液体外溢的问题，而且锌与电解液不断地发生化学反应，产生大量氢气，因此还容易爆炸，这怎么办呢？

于是，一场始于20世纪70年代的废旧电池回收运动轰轰烈烈地展开了。

Therefore, at that time, these batteries had to be removed promptly to avoid corroding the device when they were used up.

Later, people found that apart from the issue of leaked "liquid" of carbon-zinc batteries, zinc also reacted with the electrolyte to produce large amounts of hydrogen gas. And this could lead to explosions. What should people do?

Then a large-scale recycling movement for used batteries began in the 1970s.

回收废旧电池 — Recycling Used Batteries

不过，回收不如杜绝漏液根源，为了不让锌与氯化铵电解液直接接触，科学家们在锌的表面加上一层汞，使其与氯化铵电解液隔绝，由此，便诞生了碱性锌锰电池。

然而，这样做虽然解决了液体外溢的问题，但汞是一种有毒物质，这也使得废旧锌锰干电池变成了有害垃圾。

However, preventing leakage at the source is better than just recycling. To stop zinc from directly touching the ammonium chloride electrolyte, scientists added a layer of mercury as a barrier to the zinc surface. This led to the birth of alkaline zinc-manganese batteries.

This approach solves the problem of liquid leakage yet makes used batteries harmful waste, for mercury is toxic.

我们有了汞保护层，"皮肤"就不会再被损坏了。

With the mercury protective layer, our "skin" won't be damaged anymore!

不过人类好像还是不太喜欢我们啊！

But it seems that humans still don't like us!

如果这种有害垃圾大量出现，将超出环境承载极限，所以，现在干电池中的汞含量已经受到了严格的限制，很多干电池甚至可以做到完全无汞，这是怎么做到的呢？

科学家们把原先酸性的氯化铵电解液替换为碱性的氢氧化钾电解液，碱性物质不容易与锌发生析氢反应，所以，也就不需要汞的保护了。

由此，性能更加优异且环保无汞的碱锰电池便诞生了。

If this harmful waste appears in large quantities, the environment will be overwhelmed. Therefore, the mercury content in dry cells has now been strictly limited, and many dry cells can even be completely mercury-free. How is this done?

Scientists replaced the originally acidic ammonium chloride electrolyte with an alkaline potassium hydroxide electrolyte. Alkaline substances are less likely to react with zinc to produce hydrogen, so there is no need for mercury protection anymore.

As a result, this brought about the birth of a better-performing, mercury-free, and more eco-friendly alkaline manganese battery.

常见干电池的回收处理
Recycling and Treatment of Common Dry Cell

从碳锌干电池到碱性锌锰电池，再发展到今天的无汞碱锰电池的历程中，我们可以看到，电池的更新迭代与环保问题有着密不可分的关系。

From carbon-zinc dry batteries to alkaline zinc-manganese batteries, and then to today's mercury-free alkaline manganese batteries, we can see that the update and iteration of batteries are inextricably linked to environmental issues.

经过一百多年的更新迭代，我们才有了如今的完美！

It took over a hundred years of evolution for us to become so perfect!

无汞电池
mercury-free batteries

更环保
more eco-friendly

我国在2003年发布的《废旧电池污染防治技术政策》中，便明确规定了自2005年1月1日起，停止生产含汞量大于万分之一的碱性锌锰电池。

In 2003, China issued *Provention and Technology Policy for Spent Battery*. It clearly stated that the production of alkaline zinc-manganese batteries with mercury content higher than one in ten thousand would be stopped. This regulation took effect from January 1, 2005.

自此之后，由于干电池内的汞含量已经得到了有效控制，只要是一次性干电池（最常见的是碳锌电池和碱锰电池，占 80% 左右），在外壳没有损毁的情况下，用完之后，可以将其作为固体废弃物，随生活垃圾处理。它们不会污染环境，也不会进入食物链而危害人体。

Since then, the mercury content in dry cells has been effectively controlled. Now, single-use dry cells, like carbon-zinc and alkaline manganese batteries (which make up about 80%), can be safely disposed of with household refuse, as long as their casing is intact after use. They won't harm the environment or enter the food chain to harm humans.

> 如果外壳被损坏了，我们体内的有害物质还是会污染土壤！
>
> If our skin gets damaged, the harmful materials inside us could leak out and pollute the soil!

但是，如果一个地方填埋了太多的干电池，或者干电池未达到环保指标，依然有对环境造成污染的可能性。因此，对干电池进行回收再利用更有价值，并能够彻底解决电池造成的污染。

However, if too many dry cells are collected in one place, or if they do not meet environmental standards, they could still harm the environment. That's why recycling dry cells is even more important — it can completely solve the pollution problem caused by batteries.

常见干电池的回收处理技术
Recycling Techniques for Common Dry Cells

人工分选法　Manual Sorting Method

先将干电池分类成碳性电池和碱性电池，然后通过机械剖开，最后用人工方法分离出锌皮、二氧化锰（需进一步脱汞）、炭棒、塑料盖等。

First, classify dry cells into carbon batteries and alkaline batteries. Then, use machines to disassemble them and manually separate zinc casing, manganese dioxide (which needs further mercury removal), carbon rods, plastic caps, and other parts.

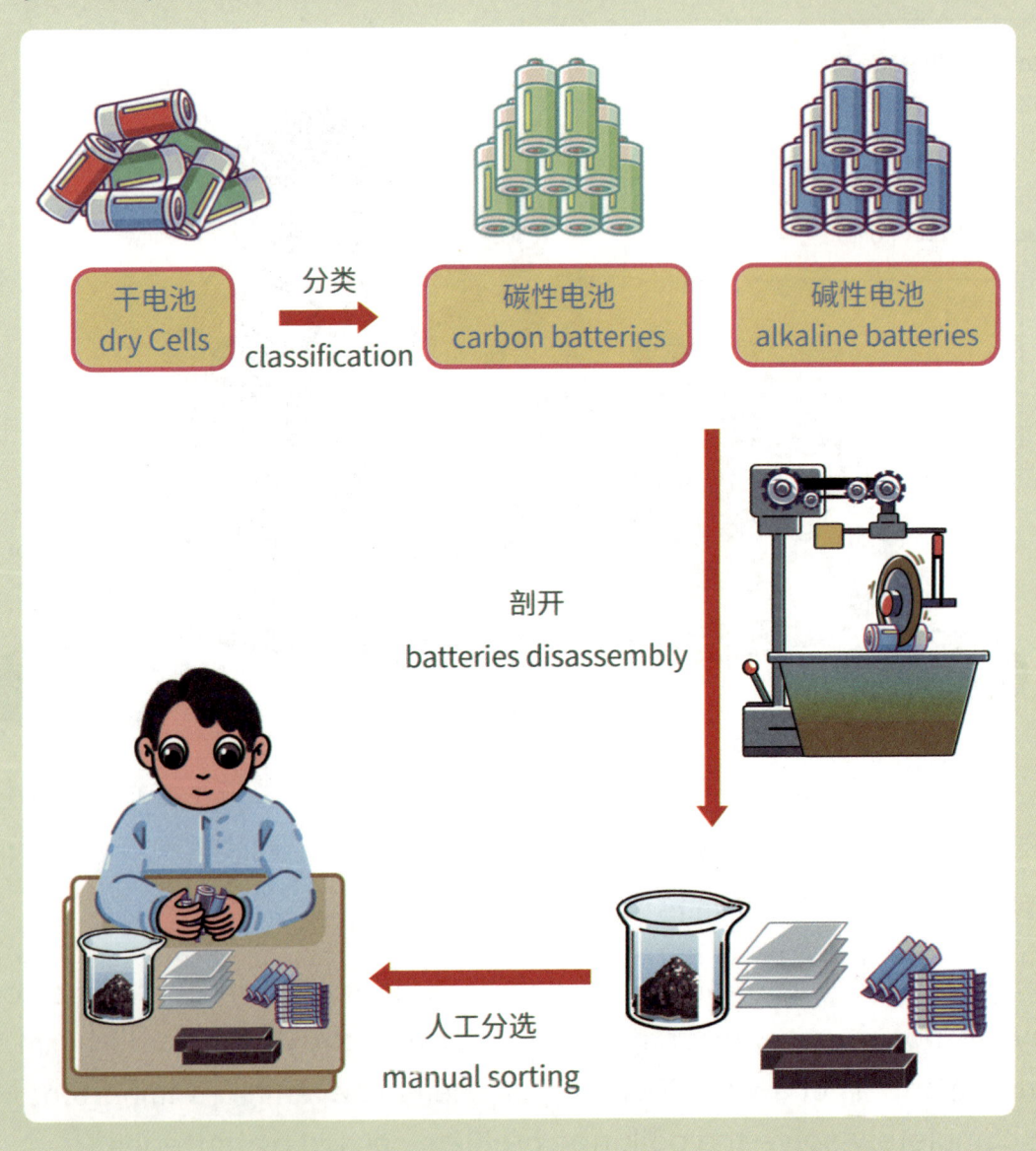

干电池 dry Cells —分类 classification→ 碳性电池 carbon batteries　碱性电池 alkaline batteries

剖开 batteries disassembly

人工分选 manual sorting

湿法回收　Hydrometallurgical Recovery

湿法回收是基于锌、二氧化锰等可溶于酸的原理，使锌锰干电池中的锌、二氧化锰与酸产生化学反应，生成可溶性盐而使其进入溶液。溶液经过净化后电解，产生金属锌和电解二氧化锰或用于生产化工产品、化肥等。

Hydrometallurgical recovery works by using the fact that zinc and manganese dioxide can dissolve in acid. In this process, the zinc and manganese dioxide in the zinc-manganese dry cell reacts with the acid, and soluble salts are formed. After purification and electrolysis, the solution is used to produce metallic zinc and electrolytic manganese dioxide or to produce fertilizers and other chemicals.

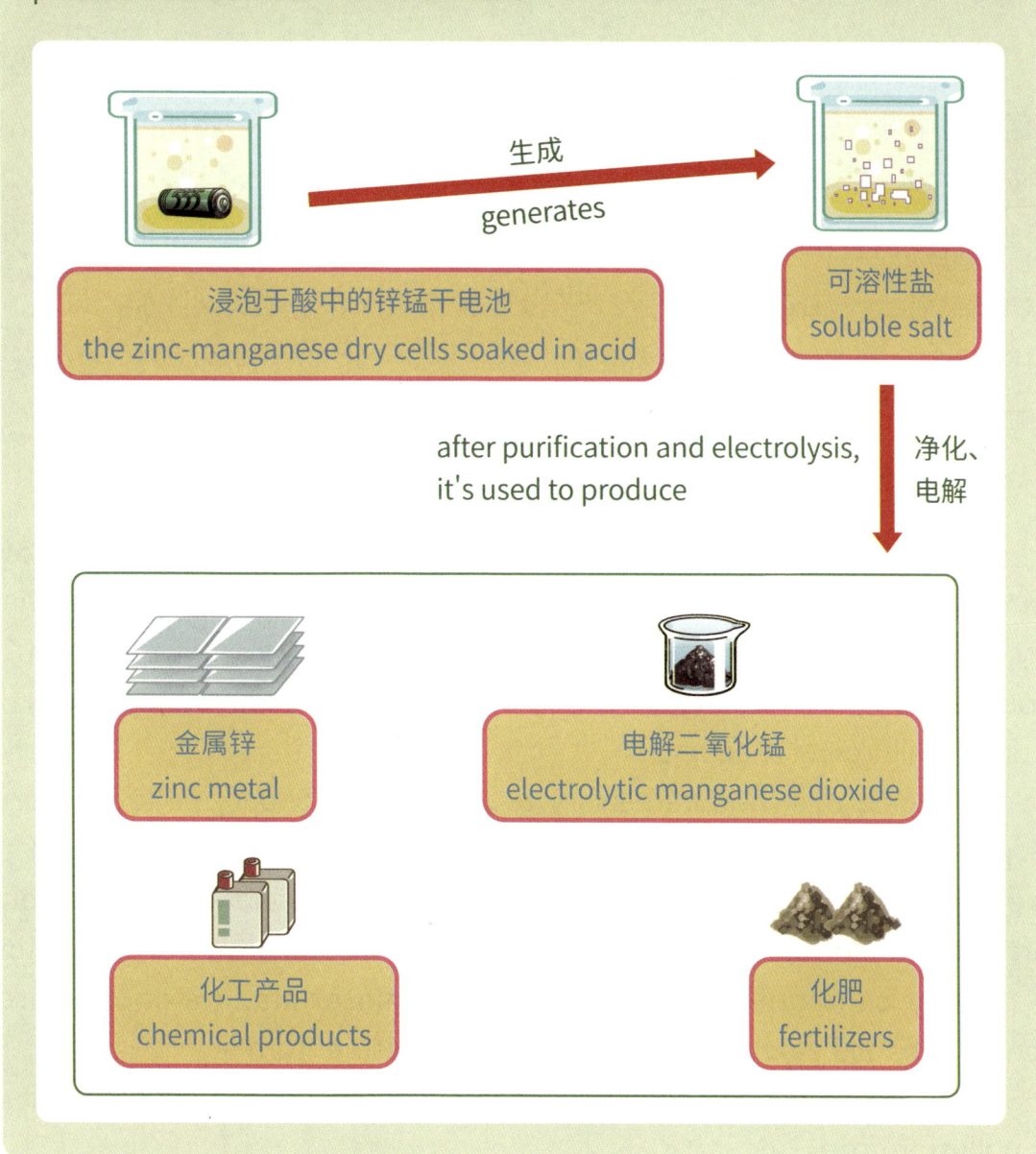

浸泡于酸中的锌锰干电池
the zinc-manganese dry cells soaked in acid

生成
generates

可溶性盐
soluble salt

净化、电解

after purification and electrolysis, it's used to produce

金属锌
zinc metal

电解二氧化锰
electrolytic manganese dioxide

化工产品
chemical products

化肥
fertilizers

火法回收　Pyrometallurgical Recovery

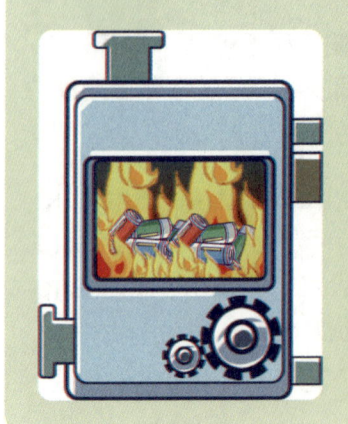

在高温下，使废旧干电池中的金属及其化合物氧化、还原、分解和挥发及冷凝。

Under high temperature, the used dry batteries and their compounds are oxidized, reduced, decomposed, volatilized and condensed.

利用上述三种方法，在经过技术处理后，可以从废旧的锌锰干电池和碱性锌锰电池中获得汞、锌、镉等金属。但还是会有遗留的有害物质，特别是湿法回收的问题更严重。

为此，需要在回收利用的过程中兼顾对二次污染的预防。可以采取的方法是：在上述三种方法中同时加入一些分选和提取步骤，延长回收加工过程，使上述方法中过于粗糙的部分更加细化，尽量恢复干电池在制成前的"原生态"，恢复铜、铁、锰、锌等的自然形态。

这样一来，在回收过程中遗留下来的溶液或者灰渣也就没有污染了。

Using the three methods mentioned above, it is possible to recover metals such as mercury, zinc, and manganese from used zinc-manganese dry cells and alkaline zinc-manganese batteries after technical processing. However, there will still be harmful residues, especially for hydrometallurgical recovery.

To tackle this, it is necessary to prevent secondary pollution during recycling. The approach is to add sorting and extraction steps to the three methods. This extends the recycling process and refines the rough parts of the methods. The ultimate goal is to restore the "original state" of the dry cells before they were made, materials like copper, iron, manganese, and zinc are recovered in their natural forms.

In this way, the solutions or ash left after the recycling process will be pollution-free.

干电池的使用常识　Tips for Using Dry Battery

装上新买的 5 号电池，怎么不亮呢？

Why won't my new AA batteries work?

　　干电池放得久了，如果受潮会漏电，继而出现"发霉"或者"生锈"的现象。这时切忌继续使用，一定要及时拆除电池，否则可能对电器本身的电路造成如短路等不良影响。

　　If dry cells sit around for too long and get wet, they may leak and cause "mold" or "rust". Don't use them when that happens. Make sure to remove the battery right away, or it could mess up the devices' circuit, maybe even causing a short circuit.

受潮后的干电池不会产生电磁辐射或核辐射，但是如镍镉电池等，会产生相当严重的重金属污染，对土壤、水源有重大危害，直接接触人体也会使人受到重金属危害，如果进入血液，甚至可能引发各种疾病。

处理完受潮的干电池，一定要及时洗手，避免重金属中毒。

Wet dry cells do not produce electromagnetic or nuclear radiation. However, batteries like nickel-cadmium ones can cause severe heavy metal pollution, harming soil and water sources. Direct contact with humans can also lead to heavy metal poisoning, and if the heavy metals enter the bloodstream, they may cause various diseases.

Always wash your hands after handling damp dry cells to avoid heavy metal poisoning.

铅酸蓄电池的环保问题
Environmental Issues of Lead-Acid Batteries

铅酸蓄电池是世界上各类电池中产量最大、用途最广的一种电池，它所消耗的铅占全球总耗铅量的82%。这种电池的技术发展比较成熟，其优、缺点都很明显。

Lead-acid batteries are the most produced and widely used type of battery in the world. They have accounted for 82% of the global lead used. This technology is mature, with clear advantages and disadvantages.

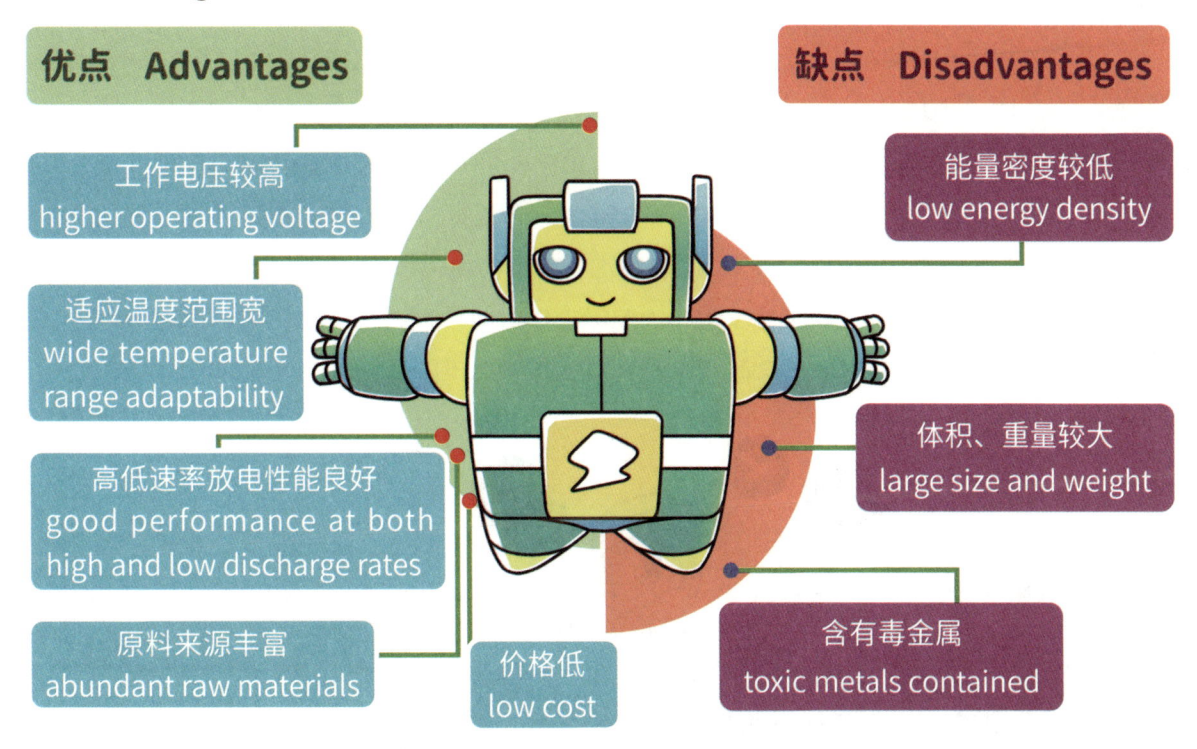

优点 Advantages
- 工作电压较高 / higher operating voltage
- 适应温度范围宽 / wide temperature range adaptability
- 高低速率放电性能良好 / good performance at both high and low discharge rates
- 原料来源丰富 / abundant raw materials
- 价格低 / low cost

缺点 Disadvantages
- 能量密度较低 / low energy density
- 体积、重量较大 / large size and weight
- 含有毒金属 / toxic metals contained

最重要的是，它体内的酸性电解液和重金属铅类材料，进入环境后会造成很大污染，对人和动物都有严重伤害。

The biggest disadvantage is that the acidic electrolyte and heavy lead materials inside can cause large pollution when released into the environment. This poses serious risks to both humans and animals.

由于铅酸蓄电池性价比高、应用领域广，暂时无法被取代，所以并未退出市场，但其报废后需要专门回收。因此，废旧铅酸蓄电池回收、处理、利用的行业技术和设备，成为重要的技术攻关项目。

自 2010 年 1 月 1 日起，我国开始实施《清洁生产标准 废铅酸蓄电池铅回收业》行业标准，标准中规定相关资质企业铅电池的回收率必须达到 95% 以上。在政策的严格要求之下，一些有先进设备的企业的回收率甚至能达到 98%，这是电池回收的一个正面例子。但与此同时，这也吸引了一些不法之徒的目光。

Because of their high cost-effectiveness and wide application, lead-acid batteries haven't been replaced yet and are still on the market. However, they require specialized recycling. Thus, the technology and equipment for recycling, processing, and reusing lead-acid batteries have become key areas for technical development.

Since January 1, 2010, China has implemented the *Cleaner Production Standard—Waste Lead-Acid Battery Recycling Industry*. It mandates that the qualified companies must achieve a lead battery recycling rate of over 95%. Under strict policy requirements, some companies with advanced equipment have achieved a recycling rate of up to 98%. This is a positive example of battery recycling. However, at the same time, this has also attracted the attention of some outlaws.

Recycling of waste lead-acid batteries

回收废旧铅酸蓄电池

由于市场上有人出高价回收铅酸蓄电池，小偷们便将目光瞄向了以铅酸蓄电池作为动力来源的电动自行车，便滋生出了"偷电瓶"这一社会乱象。

　　小偷们偷电瓶可能只是贪图一时的小利，但如果对电瓶处理不当，会对人们的健康以及生活环境带来很大影响。

Because of the high prices offered for recycling lead-acid batteries by some in the market, thieves begin stealing them from electric bikes. This results in a social problem of "battery theft".

Thieves may seek short-term profits, but improper handling of these batteries can greatly impact human health and the living environment.

铅酸蓄电池的回收现状
Current Situation of Lead-Acid Battery Recycling

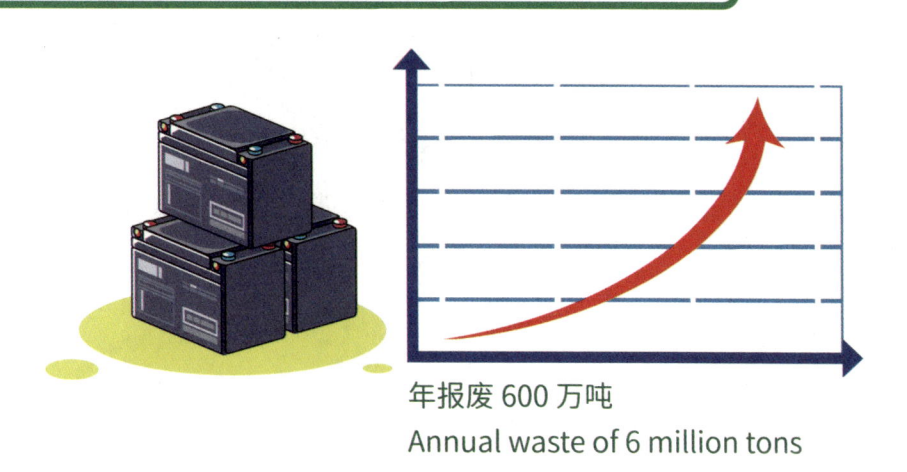

年报废 600 万吨
Annual waste of 6 million tons

　　目前，有数据显示，我国境内年报废铅酸蓄电池 600 万吨左右，且呈逐年增长的态势。虽回收率较高，但在回收、贮存、处置、利用的过程中，仍出现了大量的环境污染现象。

　　这是怎么回事呢？

Data shows that around 6 million tons of lead-acid batteries are wasted annually in China with an increasing trend. Despite high recycling rates, lots of environmental pollution occurs during the process of recycling, storage, disposal, and utilization.

What's going on here?

原来，是因为铅酸蓄电池含铅量约为74%的铅极板，在经过拆解预处理后，能获得由金属铅、铅的氧化物、铅盐及其他金属如铜、银、砷等价值不菲的材料混合而成的电池碎片。

It turns out that lead-acid batteries contain 74% lead plates. After disassembly and pretreatment, valuable materials such as metallic lead, lead oxides, lead salts, and other metals like copper, silver, and arsenic can be obtained from the battery fragments.

巨大的利益趋使一些小作坊偷偷回收拆解"电瓶"。然而，他们只需要拆卸后的铅极板，而将余下的废酸随意倾倒。这样做的后果，就是不仅回收率低，还会让溶于酸中的铅离子进入环境，最终导致水和土壤被污染。

The potential profits prompt some small workshops to secretly recycle and dismantle the lead-acid batteries. However, they only take the dismantled lead plates and carelessly dump the remaining waste acid. This results in low recovery rates and allows lead ions dissolved in the acid to leak into the environment, ultimately polluting water and soil.

铅酸蓄电池的回收处理
Recycling and Processing of Lead-Acid Battery

  小作坊的违规操作，为废旧电池的回收敲响了警钟。在做好电池回收工作的同时，人们还要保障在回收污染物时不造成二次污染。

  The illegal operations of small workshops are a wake-up call for battery recycling. While ensuring proper battery recycling, it is crucial to prevent secondary pollution during pollutant recovery.

铅酸蓄电池的回收处理技术
Recycling Techniques for Lead-acid Battery

火法处理　Pyrometallurgical Processing

采用还原熔炼，对铅酸电池进行去壳、倒酸、破碎、分选后，进行火法混合冶炼，以得到铅锑合金、精铅、软铅等材料。

Use reduction smelting to casing remoral acid draining, crushing, and sorting. Then, by pyrometallurgical smelting, materials like lead-antimony alloy, refined lead, and soft lead are gained.

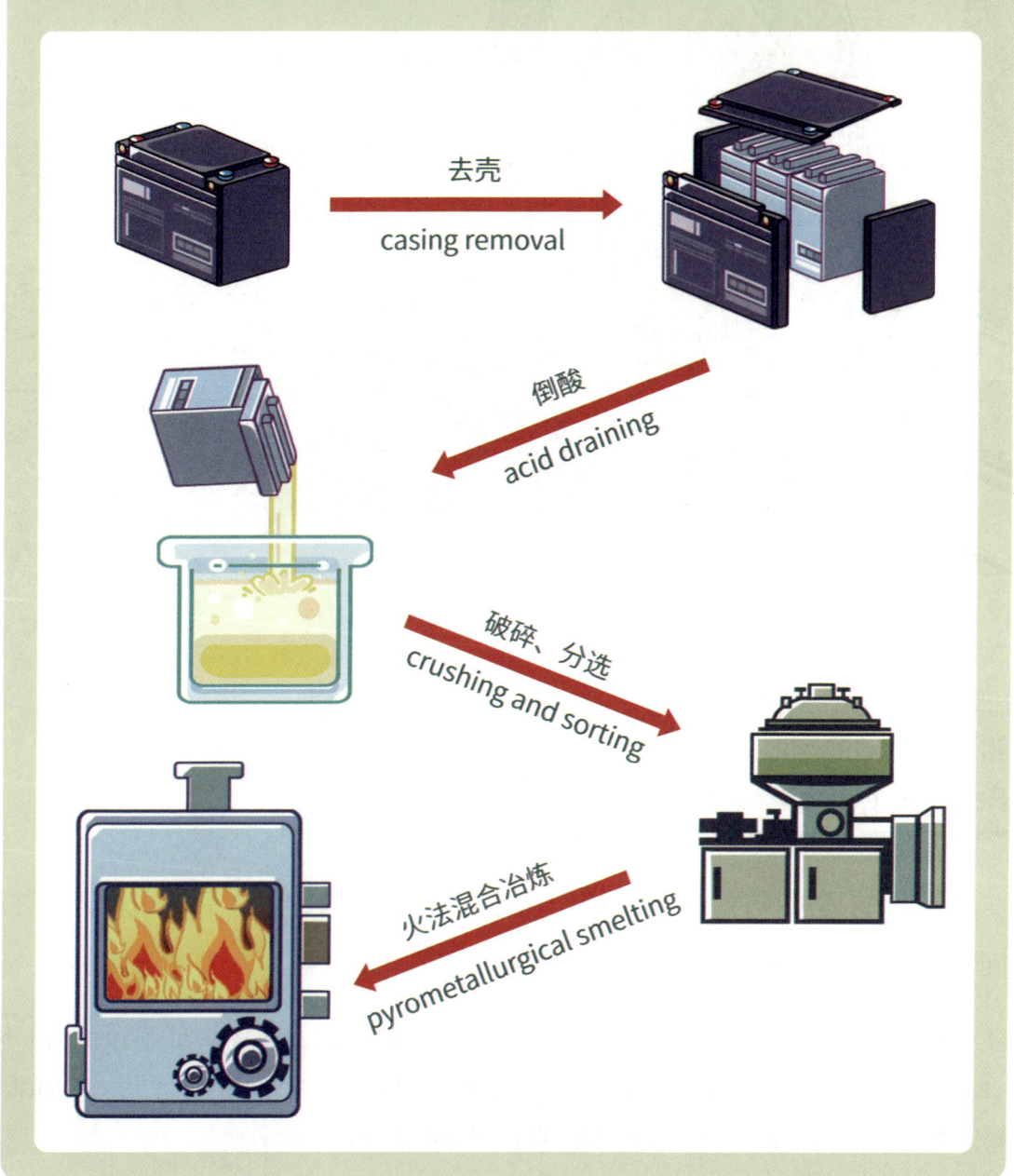

去壳　casing removal

倒酸　acid draining

破碎、分选　crushing and sorting

火法混合冶炼　pyrometallurgical smelting

湿法冶金技术　　Hydrometallurgical Recovery

湿法冶金技术，也称为"电解法"，是借助电的作用，有选择地把电池碎片中的铅化合物全部还原成金属铅。

工作原理：在溶液中加入还原剂。铅还原过程中的还原剂可用钢铁酸洗废水配置，以实现"以废治废"的目的。

此技术在冶炼过程中没有废气、废渣产生，铅的回收率可达到95%～97%。

Hydrometallurgy, also known as "electrolysis", selectively reduces lead compounds in battery fragments to metallic lead using electricity.

This is how it works: adding the reductant to the solution. The reductant, if from the waste produced during steel pickling, can put waste to better use.

No waste gas or slag is produced during smelting, with a lead recovery rate of 95%–97%.

随着人们环保意识的逐步提高，环保政策法规逐步健全，全湿法再生铅技术因其无污染的特点，将引领再生铅技术的发展趋势。

With improved environmental awareness and policies, a hydrometallurgical system emerges as the development trend for regenerated lead technology due to its non-polluting feature.

锂离子电池的环保问题
Environmental Issues of Lithium-Ion Batteries

锂离子电池的推广应用距今已经有30多年了，轨道交通、航空航天等许多国家级高精尖工程都会用到锂离子电池，电动和混合动力汽车也把它作为动力来源的最佳选择。目前，锂离子电池技术已经非常成熟和完善。

在环保方面，与其他电池相比，锂离子电池具有天然的优势。锂离子电池的包装材料不含镉、铅、汞等有害重金属，在生产和使用过程中不产生任何污染物，不会对环境造成重金属污染，其回收过程中产生的水土污染问题相对容易解决。

但是,用完的锂离子电池一样需要回收。这是因为，它的加工要用到钴酸锂、铜、铝、镍等，要是随便丢弃,还是可能对环境造成一定的影响。

Lithium-ion batteries have been around for over 30 years and are used in many advanced national projects such as rail transportation and aerospace. They are also the optimal power source for electric and hybrid vehicles. Currently, lithium-ion battery technology is highly mature and refined.

In terms of environmental protection, lithium-ion batteries have natural advantages over other batteries. Their packaging materials do not contain harmful heavy metals like cadmium, lead, and mercury, and they do not produce any pollutants during production and use, avoiding heavy metal pollution to the environment. Water and soil pollution issues in recycling are relatively easy to solve.

However, used lithium-ion batteries still call for recycling due to lithium cobalt oxide, copper, aluminum, and nickel involved in their processing. They can harm the environment if discarded carelessly.

锂离子电池拥有快充功能，能随用随充，并且没有记忆效应，不会额外损耗电池。相较于铅酸蓄电池而言，它的使用时间更长，无须频繁更换，更省心。

　　随着新能源汽车的发展，锂离子电池由于其性能优异，和其他电池相比，产业规模更为庞大。尤其在锂离子电池被作为电动汽车的首选动力电源后，锂离子电池的数量更加庞大。据最新统计，我国 2024 年锂离子电池产量累计值为 2 945 707 只。这么多的锂离子电池，会不会造成回收困难的问题呢？

Lithium-ion batteries support fast charging and can be recharged anytime without causing a memory effect, avoiding extra battery wear. Compared to lead-acid batteries, they last longer and require less frequent replacement, making them more convenient.

As new energy vehicles continue to thrive, lithium-ion batteries have driven their own industry to a much larger scale for better performance than other types. Especially since lithium-ion batteries became the optimal power source for electric vehicles, their numbers have increased significantly. According to the latest statics, the cumulative output of lithium-ion batteries in China in 2024 is 2 945 707. Will so many lithium-ion batteries cause recycling diffculties?

Used Lithium-ion Battery Recycling

废旧锂离子电池回收站

回收难, 不要挤！
Recycling is hard, don't push!

Please recycle me!

求回收

请回收我！
Please recycle me!

锂离子电池的回收处理
Recycling and Processing of Lithium-ion Batteries

锂离子电池的主要组成部分包括外壳、电解液、正极材料、负极材料、黏合剂、铜箔和锡箔等，具有很高的回收价值。

有专家表示，经过回收处理后的锂离子电池的材料，往往会随着制造和精炼的每一个额外循环而变得更加纯净。

The main components of lithium-ion batteries include the casing, electrolyte, cathode materials, anode materials, adhesives, copper foil, and tin foil, all of which have high recycling value.

Experts say that materials from recycled lithium-ion batteries often become purer with each additional cycle of manufacturing and refining.

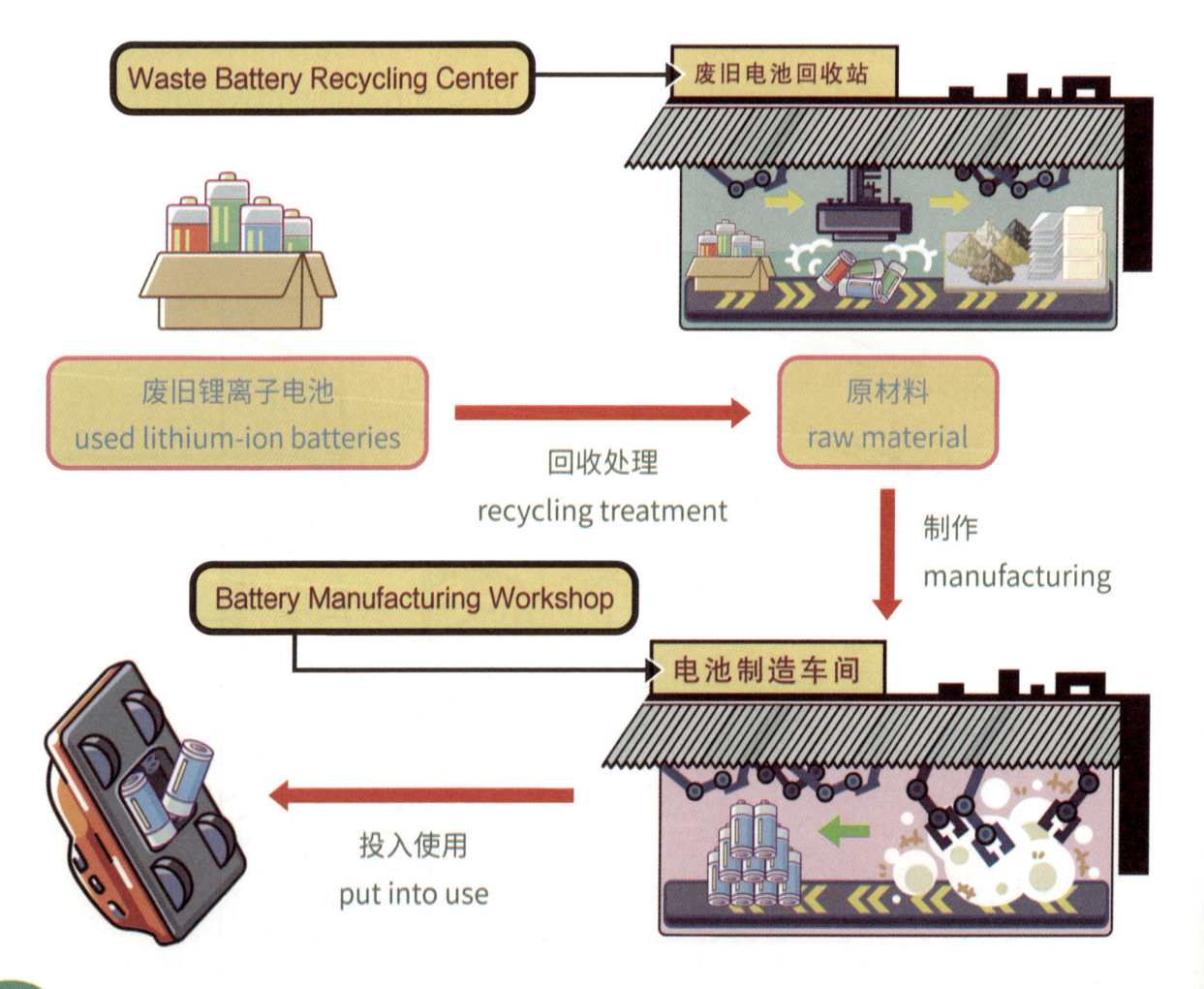

Waste Battery Recycling Center — 废旧电池回收站

废旧锂离子电池 / used lithium-ion batteries

回收处理 / recycling treatment

原材料 / raw material

制作 / manufacturing

Battery Manufacturing Workshop — 电池制造车间

投入使用 / put into use

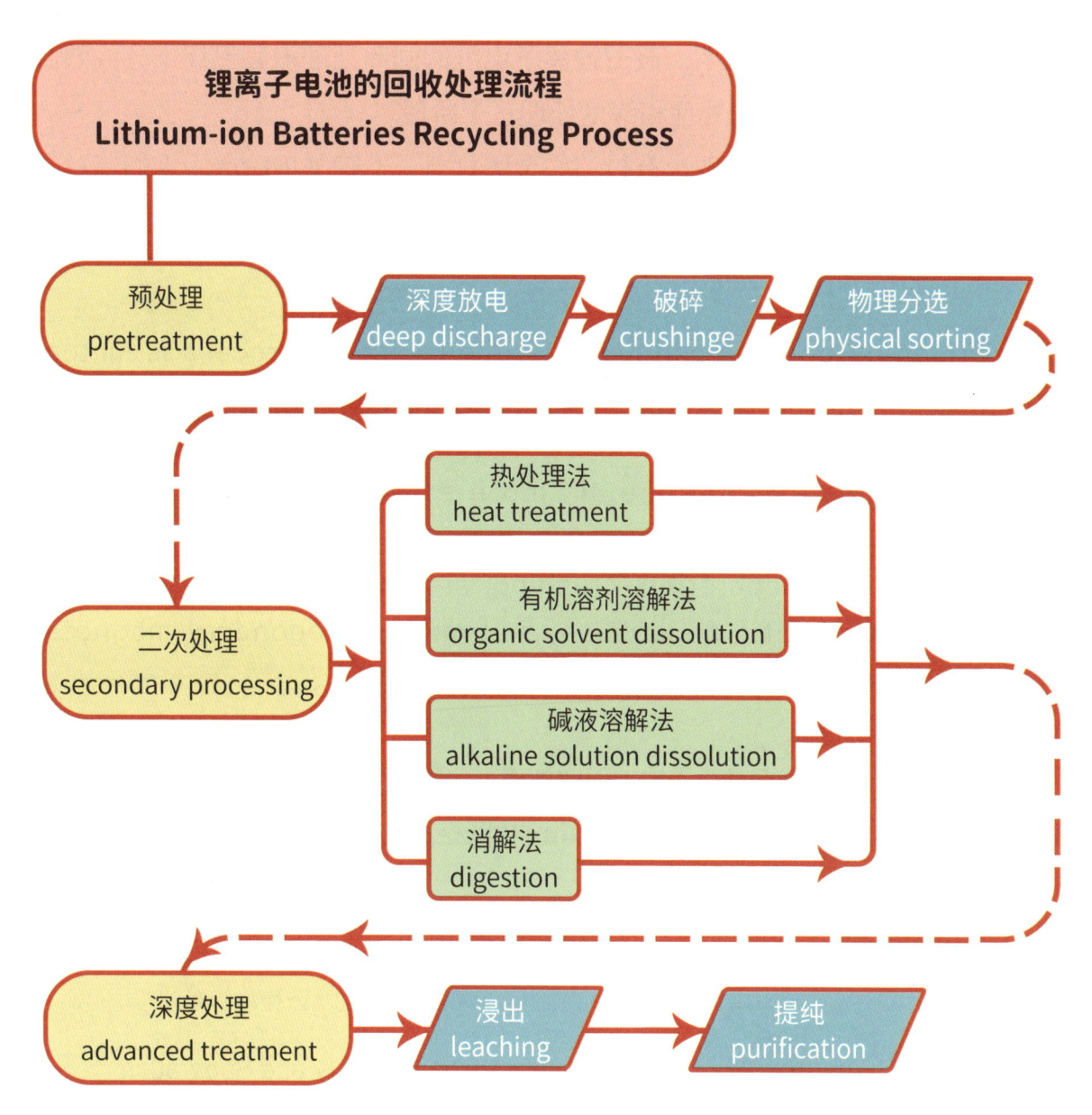

锂离子电池的回收处理流程
Lithium-ion Batteries Recycling Process

```
预处理 pretreatment → 深度放电 deep discharge → 破碎 crushinge → 物理分选 physical sorting
                ↓
二次处理 secondary processing → 热处理法 heat treatment
                            → 有机溶剂溶解法 organic solvent dissolution
                            → 碱液溶解法 alkaline solution dissolution
                            → 消解法 digestion
                ↓
深度处理 advanced treatment → 浸出 leaching → 提纯 purification
```

回收的过程不仅不会影响电池的性能,还可以提高质量,从而生产出更好的电池。

这是一个令人惊讶的结果。最近的科学研究已经证实——回收的电池不仅性能与新电池一样好,而且使用寿命更长,充电速度更快。

The recycling process does not weaken the performance, yet improves the quality, thus producing better batteries.

This is a surprising result. However, recent scientific research has confirmed that recycled batteries perform just as well as new ones with a longer lifespan and faster charging speed.

在政策和市场的推动下，废旧锂离子电池的回收处理流程已经十分成熟，主要划分为预处理、二次处理和深度处理三个阶段。

为了能够更有效地提取出有价值的金属材料，在预处理阶段，人们还总结出了八道具体工序，依次分别是：拆解分离、破碎、过筛、碾压粉碎、一次磁选分离、球磨粉碎、二次磁选分离和分类。

Driven by policies and markets, the recycling process is mature and divided into three stages: pretreatment, secondary processing, and advanced treatment.

To extract valuable metals more effectively, eight specific procedures have been summarized in the pretreatment stages: disassembly and separation, crushing, sieving, rolling and crushing, preliminary magnetic separation, ball milling, secondary magnetic separation, and classification.

废旧锂离子电池回收处理——预处理工序
Pretreatment Steps for Recycling Waste Lithium-Ion Batteries

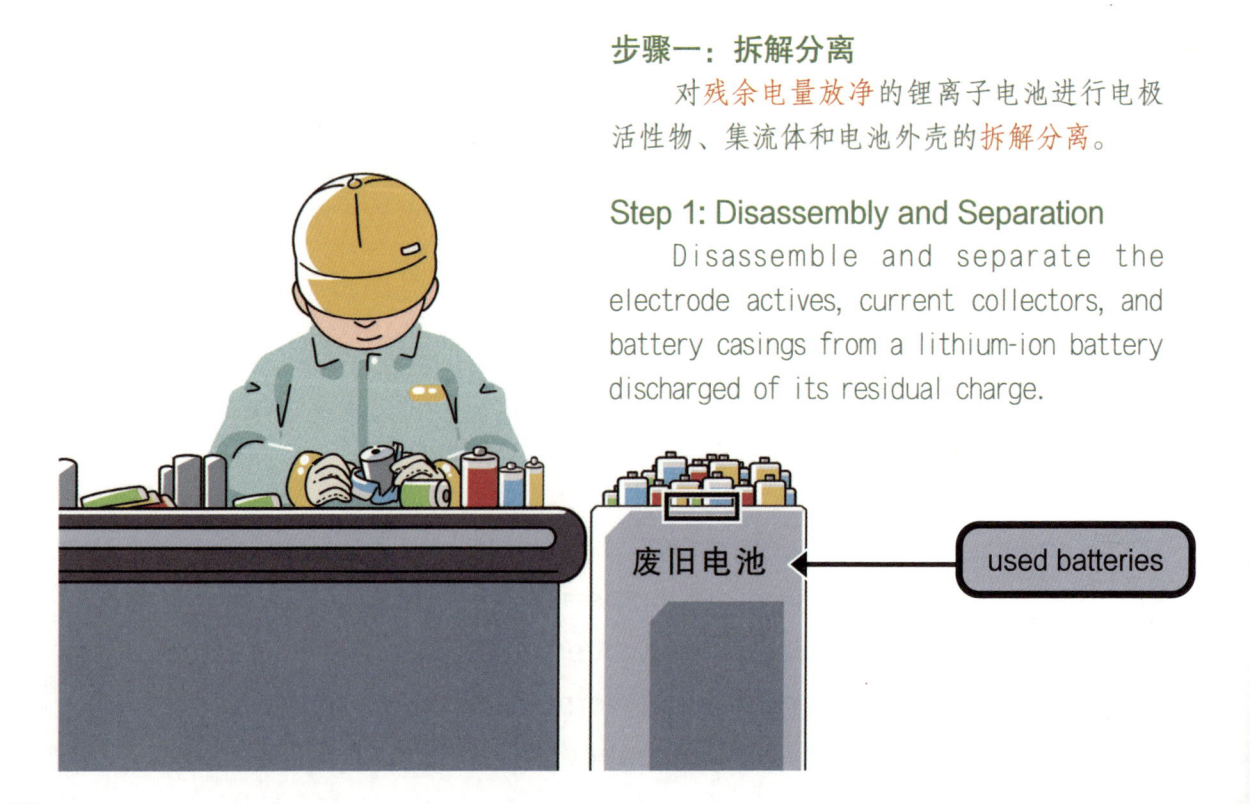

步骤一：拆解分离
对残余电量放净的锂离子电池进行电极活性物、集流体和电池外壳的拆解分离。

Step 1: Disassembly and Separation
Disassemble and separate the electrode actives, current collectors, and battery casings from a lithium-ion battery discharged of its residual charge.

废旧电池 ← used batteries

步骤二：破碎

将完成拆解的电极活性物、集流体和电池外壳用破碎机进行破碎。

Step 2: Crushing

Use a crusher to crush the disassembled electrode actives, current collectors, and battery cases.

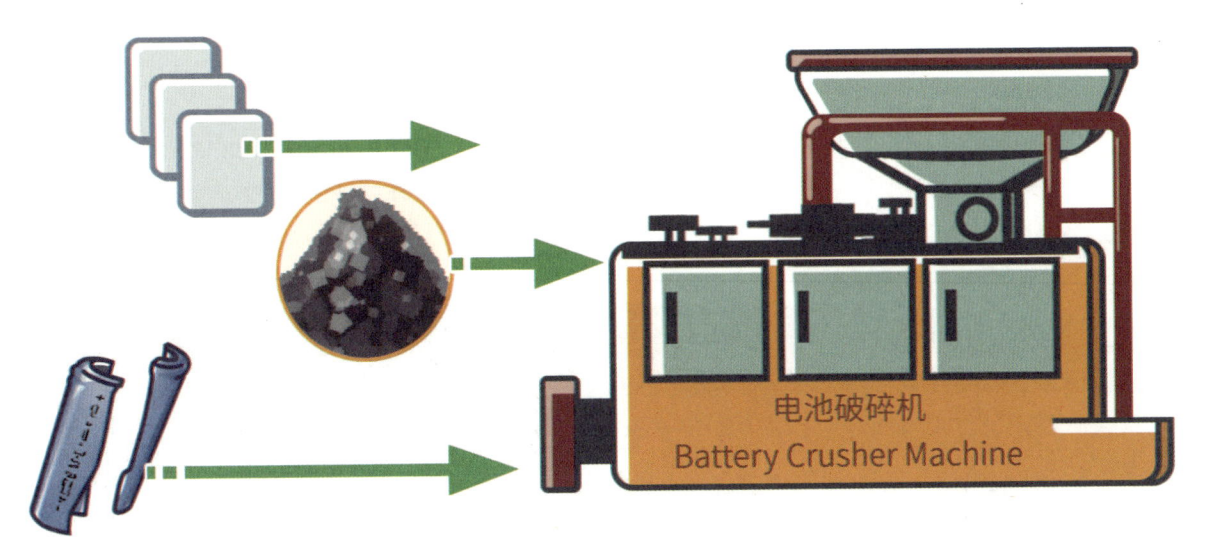

电池破碎机
Battery Crusher Machine

电池小知识 Battery tip

锂离子电池破碎设备能把电池中含有的钴、镍、锂等金属分选出来，通过科学系统的高效回收，回收率能达到98%以上，在很大程度上补给了动力电池的生产，缓解了锂资源的紧张情况。

Lithium-ion battery crushing equipment can separate cobalt, nickel, lithium, and other metal from batteries. Through systematic recycling, the recovery rate can reach over 98%, significantly supplying the production of power batteries and easing the lithium shortage.

步骤三：过筛

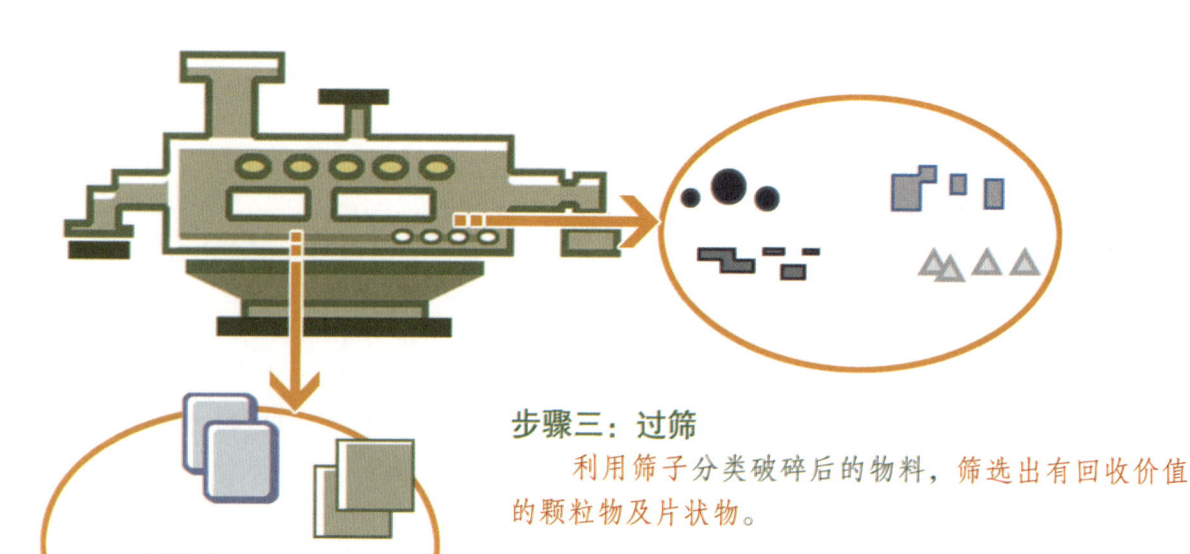

利用筛子分类破碎后的物料，筛选出有回收价值的颗粒物及片状物。

Step 3: Sieving

Use a sieve to classify crushed materials, separating granules and flakes with recycling value.

> **小贴士 Tips**
>
> 利用筛子筛选出的颗粒物的主要成分为金属氧化物、金属部件、碳素材料和碳类化合物；筛选出的片状物主要成分为隔膜、有机物和钢壳。
>
> The main components of sieved granules are metal oxides, metal parts, carbon materials, and carbon compounds. The sieved flakes consist of separators, organic substances, and steel shells.

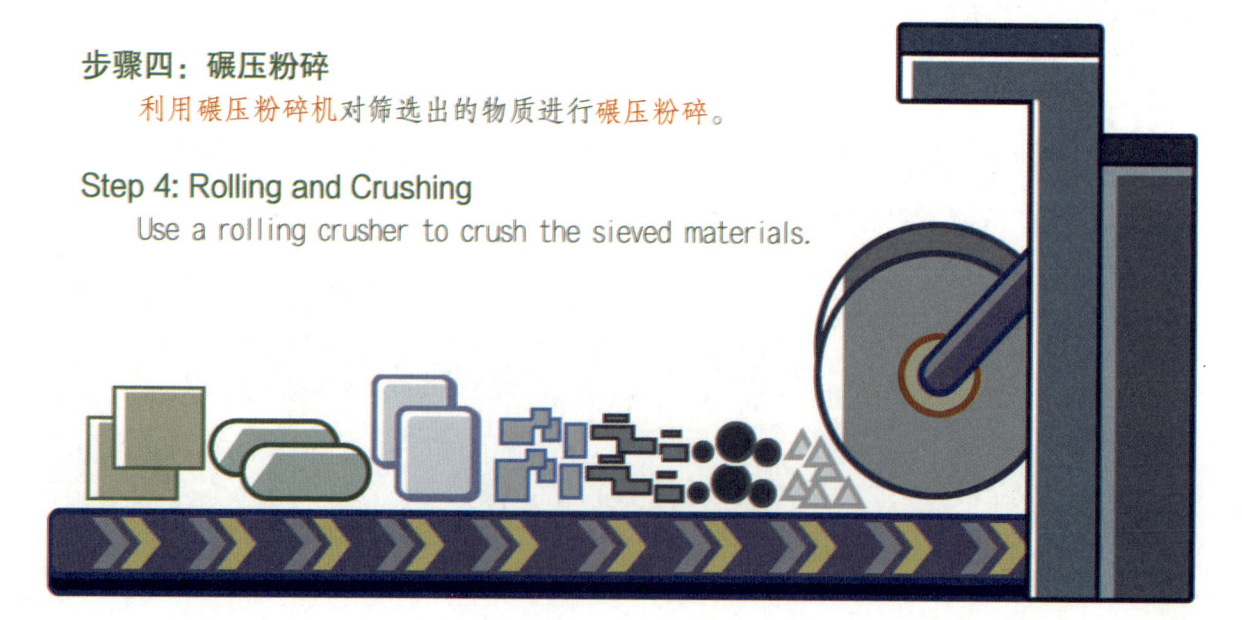

步骤四：碾压粉碎

利用碾压粉碎机对筛选出的物质进行碾压粉碎。

Step 4: Rolling and Crushing

Use a rolling crusher to crush the sieved materials.

步骤五：初步磁选分离

通过磁选机对碾压粉碎后的物料进行磁选，使不同磁导率的金属颗粒被筛选出来并进行初步分类收集。

Step 5: Preliminary Magnetic Separation

Use a magnetic separator to sort crushed materials by magnetic separation. In this way, metal particles with different magnetic conductivity are separated and preliminarily classified for collection.

magnetic separator

磁选机

步骤六：球磨粉碎

利用球磨机依次对经过初步分类收集的金属颗粒物进行球磨，使金属颗粒物更细化。

Step 6: Ball Milling

Use a ball mill to further refine metal particles collected from preliminary classification.

球磨机

ball mill

secondary magnetic separator

二次磁选机

步骤七：二次磁选分离

利用二次磁选机对经过球磨机球磨后的直径更小的颗粒进行磁选，使不同磁导率的金属颗粒被更加精细地筛选出来。

Step 7: Secondary Magnetic Separation

Use a secondary magnetic separator to sort smaller particles from ball milling, finely sorting metal particles by magnetic conductivity.

步骤八：分类

对二次磁选分离的金属颗粒进行分类，可准确得到不同种类的金属颗粒。

Step 8: Classification

Classify metal particles from secondary magnetic separation to precisely get different types of metal particles.

个人应该怎么处理废旧电池？
How Should Ordinary People Handle Used Batteries?

随着电子产品、家电种类越来越丰富，电池作为电力的贮存装置，废旧电池的数量以及种类也在不断增加。

大家现在知道，废旧电池中含有许多重金属及电解质溶液，会对人体和生态环境造成危害，并且已经了解电池在"退役"之后，工厂是如何处理它们的。那么，在日常生活中，我们能够为它们做些什么呢？

With more diverse electronics and appliances that rely on batteries as power storage devices, the number and variety of waste batteries are also growing.

It is well-known that used batteries contain many heavy metals and electrolyte solutions. Those materials harm humans and the environment. After learning how factories handle "retired" batteries, what can we do to deal with them in our daily lives?

废旧电池回收箱
Used Battery Recycling Box

你可以选择重复利用，例如玩具车用过的电池，直接扔掉比较浪费，可以将它们装进电视遥控器里进行二次利用。因为遥控器不需要太高的电压，即便是废旧电池也够遥控器用一段时间了。

You can choose to reuse batteries, like those from toy cars, as throwing them away would be wasteful. Given the little voltage required, they can be reused in the TV remote control for quite some time.

如果你是一位艺术爱好者，还可以将废旧电池收集起来，把它们组装成不同造型的装饰品。

If you're an artist, you can collect used batteries and assemble them into various decorations.

如果你实在是懒得折腾的话，应该做好垃圾分类，将它们单独丢弃到专门回收电池的垃圾箱里去。

If you prefer not to bother, practice waste sorting and dispose of them in dedicated battery recycling boxes.

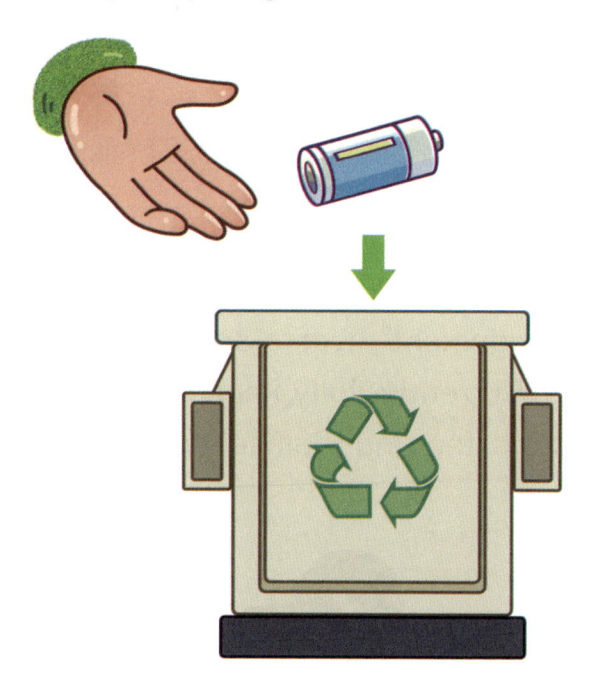

自然环境是人类社会赖以生存的前提和基础，是人类创造活动的平台。我们必须重视对环境的保护，从个人出发、从小事做起，为人类可持续发展创造足够的空间。

The natural environment is the foundation for human survival and a platform for creation. We must prioritize protecting it, start small, and act now, making human development more sustainable.

废旧电池属于什么垃圾？
What Type of Used Are Waste Batteries?

自我国部分城市生活垃圾管理条例正式实行，垃圾分类已成为居民的"必修课"。目前，垃圾分类的全面实施还存在一定的难度，大家在丢垃圾时，也会纠结到底该投到哪个垃圾桶。

Since some cities in China implemented the "Regulations on Domestic Waste Management", waste sorting has become a "mandatory course" for residents. Currently, fully implementing waste sorting remains challenging, as people often find themselves unsure of which box to cast for their waste.

电池中的重金属危害那么大，是不是属于有害垃圾呢？无汞化的碱性锌锰电池是否可以随其他垃圾一起扔掉？否则又该丢到哪里呢？

Given the heavy metals of great hazard in batteries, are they considered hazardous waste? Alkaline zinc-manganese batteries are mercury-free, can they be discarded with other waste? Or can you guess where they should go?

家庭生活中比较常见的一次性玩具电池、家电的遥控电池等，使用的是在市场上出售的1号、5号、7号干电池，它们一般为碱性锌锰电池，都可以被投放到"其他垃圾"桶中。

二次电池(俗称"充电电池"，包括镍镉、镍氢、锂电池与铅酸蓄电池)、纽扣电池中均含有重金属，属于危险废弃物，对环境危害大，需要专门回收处置，因此属于有害垃圾，要投放到"有害垃圾"桶中。

Residual Waste box: disposable batteries in toy, remote control and other household items, or dry batteries such as D batteries, AA batteries, and AAA batteries sold on the market.

Hazardous Waste box: secondary batteries (commonly known as "rechargeable batteries", like nickel-cadmium, nickel-hydride, lithium batteries, and lead-acid batteries) and button cells. Containing heavy metals of hazardous wastes, those batteries pose a great danger to the environment and require specialized recycling accordingly.

图书在版编目（CIP）数据

电池大环保：汉英对照 / 马建民主编；咪柯文化绘图；廖敏译. -- 成都：成都电子科大出版社，2025.

1. -- ISBN 978-7-5770-1488-3

Ⅰ. TM911-49

中国国家版本馆 CIP 数据核字第2025WW5571号

电池大环保（中英对照版）
DIANCHI DA HUANBAO（ZHONG-YING DUIZHAO BAN）

马建民　主编　咪柯文化　绘　廖　敏　译

策划编辑	谢忠明　段　勇
责任编辑	赵倩莹
责任校对	蒋　伊
责任印制	段晓静

出版发行	电子科技大学出版社
	成都市一环路东一段159号电子信息产业大厦九楼　邮编610051
主　页	www.uestcp.com.cn
服务电话	028-83203399
邮购电话	028-83201495
印　　刷	成都久之印刷有限公司
成品尺寸	185 mm×260 mm
印　张	6
字　数	114千字
版　次	2025年1月第1版
印　次	2025年1月第1次印刷
书　号	ISBN 978-7-5770-1488-3
定　价	66.00元

版权所有，侵权必究